KV-372-279

# Micros in Process and
Product Control

# Micros in Process and Product Control

## A.A. Berk

**COLLINS**
8 Grafton Street, London W1

Collins Professional and Technical Books
William Collins Sons & Co. Ltd
8 Grafton Street, London W1X 3LA

First published in Great Britain by
Collins Professional and Technical Books 1986

Copyright © A.A. Berk 1986

*British Library Cataloguing in Publication Data*
Berk, A.A.
Micros in process and product control.
1. Process control—Data processing
2. Microcomputers
I. Title
670.42'7        TS156.8

ISBN 0-00-383296-1

Photoset in North Wales by
Derek Doyle & Associates, Mold, Clwyd
Printed and bound in Great Britain by
Mackays of Chatham, Kent

# Contents

# Foreword

There is often a gap between the users of a given technology, and the experts who design and apply it. This is probably more true of the microelectronics field than any other, and it can hamper the adoption of these vital devices even in fields where it is obvious that they are required. This book aims to bridge that gap, and provide a solid background in control devices and their peripherals for anyone who needs to use controllers in a given project. It provides information on the controllers themselves, in terms of the correct level to choose in a given situation. The book also describes common peripherals such as sensors and actuators, and aims to give the reader enough information to perform the initial systems analysis which points the project in the correct direction. The approach is to describe examples of control projects throughout, for use by the reader as templates.

The level of explanation adopted is practical without being too electronically detailed. Design and discussion are taken to the point where an electronic engineer can be brought in to design the system. This gives the non-specialist valuable assistance in dealing with experts in the otherwise mysterious field of control. It also provides ideas and some of the essential background data to consider, perhaps, new ways of viewing the control of products and processes.

The examples used throughout are from my own experience of the ways in which microprocessors and electronics have been used or proposed in manufacturing industries. In many ways the book arises from a need which has become apparent in the personnel with whom I have dealt in those industries. In general, it has been found time and time again that the more knowledgeable those people are, the more efficient is the introduction of microelectronics to their working environment.

The book is suitable for managers and other personnel with a mechanical, electrical or other engineering/technical background who will need to use microelectronics in their industry. It will also be of assistance to programmers who wish to extend their activities into control, but who have worked, until now, in data processing for instance. Though there is a fair amount of technical explanation here, the background material will help them to gain a more low level and hardware-orientated understanding of the technology, which is crucial to the successful programming of controllers.

I should like to thank several people who have helped me considerably. Firstly, my wife and family who have shown their usual interest and understanding during the associated periods of continuous work at all hours. Secondly, to my father who has, as so often in the past, spent many hours reading and commenting upon the material in the text. My thanks are also due to those companies who have supplied photographs for inclusion within the text. Finally, I should like to thank the editorial staff of Collins, whose professionalism has, as always, made the task of preparing the book for publication both pleasurable and rewarding.

Dr. A. A. Berk

Chapter One
# General Control
# Microelectronics

## INTRODUCTION

This chapter introduces a number of aspects of control in processes and products within the manufacturing industry. In general, the book deals with the control (or automation) of both the processes used by the manufacturing industry and the products, where appropriate, being manufactured. As will be seen, the areas encompassed are broad and range across the spectrum of engineering endeavour. For instance, it is not difficult to find a single microelectronic controller which could be used in a plastic moulding plant or a chess playing machine. As will be seen, economics may decide that one of these applications is less efficient than the other. It is also true that the software (programming) within a given piece of computer hardware is the main deciding factor as to the final application – the hardware is often said to be a pawn in the hands of the programmer.

We will start the present chapter by agreeing upon some definitions of the most common terms used in the book. Controllers in manufacturing situations are then described according to a type of classification which will allow control situations and hardware to be seen in context. This inevitably includes a look at simple systems analysis for a process or general control situation. Different types of controller are described in black box form. The internal design of controllers in general is examined in Chapter 2.

There are two appendices which will be of assistance in reading this chapter, as well as the rest of the book. The first gives a roundup of the essentials of binary numbers and associated information. The second appendix gives company names and addresses as contacts in the field of microelectronics and control, along with a short explanation of their products.

## SOME SIMPLE DEFINITIONS

The word 'controller' will normally mean any computer circuit which controls the outside world. This is very wide, as it can be argued that a large computer installation (a mainframe, for instance) is controlling the outside world via a screen and keyboard when an operator is using the machine. Similarly, a single integrated circuit will be found within a musical door chime, performing the job of controlling a loudspeaker to produce a noise when someone presses a bell-push. This just about spans the complete gamut of computing devices, and as such means that the word 'controller' can apply in almost any computer situation. In addition, it is possible for non-computer circuits to be controllers too. In general, you will naturally pick up the meaning of the word 'controller' in context as we proceed, but you should remain aware of the complexity of this term.

The word 'computer' is much maligned and over-used. In this book it will have one of several meanings. Since we are concerned here with 'intelligent' electronics, i.e. electronics which can be programmed, our basic hardware will inevitably be based around a microprocessing unit (MPU), or microprocessor. This is an electronic circuit which is capable of executing a set of instructions supplied to it by some external electronics, and written by a programmer.

The word 'computer' will be used to describe a circuit containing an MPU. We will also be concerned, later, with larger computers (known as development systems) which are used, essentially, in the design of MPU-based controllers. As it happens, most of these larger machines will also be based around MPUs.

To place the MPU in its position within the context of engineering, you should consider it as the centre of a network of electronic circuits which are normally devoted to supplying it with a program, fixed data, memory to store its variable data and with interfaces to the outside world. You should further consider such a network to be the most basic and general kind of electronic circuit possible. The general point is that by a change in the program controlling the MPU, the network can be made to emulate any kind of electronic circuit – including analogue circuits, controllers, electronic filters, and so on. It is true that some extra hardware may be needed to complete the system, and that it may be too slow or uneconomic to replace some circuits with an MPU-based device, but this is the art of choosing the correct level of machine for a given application. This book aims to provide the criteria for choice within the engineering field of control – other books are devoted to the use of the MPU in other fields.

Hardware and software are words which are used freely in the industry – most people know the meanings, i.e. that hardware describes the electronics and mechanics of a project while software describes any

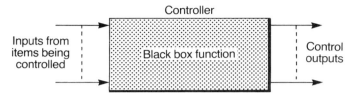

*Fig. 1.1* Black box approach.

computer programming which may be involved. As a corruption of these two terms, one may talk of a 'soft solution' to a problem, for instance. This simply means that the solution lies with a standard piece of hardware, and the problem is solved by writing the computer's program in a special manner. There are often two solutions to a given part of a project – hard and soft. A 'hard solution' is normally a solution which depends upon the specific design of a new piece of special hardware.

A black box approach, as in Figure 1.1, will mean an approach which connects a set of inputs and outputs (I/O) to a box and defines how the box acts on these I/O lines. The black box approach does not try to explain the way in which such a box works internally. This is a favourite approach in the analysis of a problem. One can therefore ask the question 'which type of controller (black box) will act in the following way?'. This approach allows the *function* of the problem to be examined and defined without concern for the internal electronics of the black box which solves the problem. It is a good way to approach the setting up of a specification without having to understand microtechnology. In many cases the details of the electronics are standard once the problem is defined.

A certain confusion exists in the use of the word 'microelectronics'. It can be argued that micro circuits which contain many electronic components shrunk to the size of an integrated circuit, or silicon chip, are microelectronic. Others use the term exclusively to mean circuits employing or at least directly related to MPUs. We will tend towards the latter definition in this book, as it is becoming widespread. A microelectronic solution to a problem is fast becoming synonymous with a computer-based solution.

The words 'integrated circuit', 'IC', 'silicon chip', 'chip', and even in some books 'bug' will all be synonymous. These are terms used to describe the type of device where many electronic components are formed onto a small piece of semiconducting material in order to integrate a number of electronic functions into one device.

There are two important comments to bear in mind at this point. The first is that an electronic solution not employing MPUs and programming may be an economic and convenient solution to a control problem and this must be borne in mind throughout. Secondly, the area of the dedicated custom-made integrated circuit for a given job is becoming more and more widespread in the industry. This can lead to integrated circuits of the complexity of an MPU with rather different characteristics. It is fair, in many ways, to use the term microelectronics to include this field too.

'Process control' will generally mean the control of a manufacturing process, and 'product control' will generally apply to the electronics built in to a given manufactured product. For instance a washing machine controller provides an example of product control. The design of a device to control the pressing of steel sheet into side panels for the washing machine, or the moulding of its plastic parts is a problem in process control. A good example of a case where these two definitions overlap is in the case of a controller within the machine which performs the plastic moulding of the lining of a refrigerator. It is process control to the refrigerator manufacturer, but product control to the manufacturer of the moulding machine itself!

Input/output is a term commonly used to describe electronic lines which connect a given black box to devices which someone within the black box might refer to as the outside world. For instance, a door chime controller would have electrical lines connecting to a loudspeaker, to a bell push, and perhaps to a rotary switch which selects the tune to be played. These are referred to as I/O lines for short. In many ways, the key to an analysis of any given control situation is to discover the number and character of all the I/O lines needed.

As a further example, we might consider the control of a washing machine. Choosing a suitable controller depends upon price – it is important in a high volume device to keep the cost of each part to a minimum. The first question is how many I/O lines must the controller have. This is easily estimated as follows. Some lines are needed to read a rotary switch and a couple of special on/off switches on the front panel. Two AC electric motors are to be controlled along with a couple of fluid valves. Four emergency sensors are needed, say, for such parameters as over-temperature, water supply failure, etc. This is shown in Figure 1.2. Of course, in practice there will be more items to be controlled, but this illustrates a first consideration of the problem.

It should be noted that in talking about the number of lines required for a given set of functions, the electrical returns are often not mentioned. Single I/O lines will normally act with respect to a common return which may be the earth or ground of a power supply. Alternatively, twin lines may be needed in some circumstances. Example circuits will be seen later.

Given that a 16-position rotary switch can be read using just four lines (see Appendix 1) the total number of I/O lines required for this washing machine will be around 14, which is below a magic number – 16 – and this can be catered for by a prodigious number of 'single-chip' computers. Such chips contain the MPU itself along with all the peripheral circuits required to turn it into a complete computer, with the program itself formed onto the chip during manufacture. I/O lines are also included – often a minimum of 16. Output lines normally require some external electronic interfacing (I/F

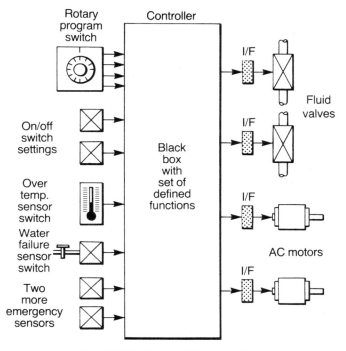

*Fig. 1.2* Washing machine controller.

as shown in Figure 1.2) to change their low voltage levels into power switching for mains or other powerful devices, but such electronics is cheap and readily available. In general, the inputs also require some interfacing, as the outside world does not normally run on the same voltages as the computer circuits.

This is an example of a quick, simple view of the systems analysis of a given control situation; it almost always starts with the question 'how much I/O does it need?'. It is generally considered that the way in which that I/O is controlled, i.e. the function of the device – is a straightforward matter of program writing. Of course, this is a simplification as the programming of some types of control situation can be quite complex within the required parameters. Sometimes, the programmer will demand large changes to the electronics in order to make his task easier. This is an important characteristic of control electronics – it is a complete weld of hardware and software.

One of the main differences between hardware and software is that shaving a few components off a given controller will save a lot of money in a product manufactured in high volume. Shaving a few lines off a program, or even doubling the size of a program will often have no effect on the manufacturer's unit cost. However, the amortised development cost may be affected significantly by large changes in software.

The overall system designer has to decide upon the correct direction of the project at the very start by constructing a compromise based on all the factors. For instance, a path which may be cost effective in a given volume manufacturing situation may be inappropriate in a defence contract where reliability and operational effectiveness may be a prime requirement, and cost of less importance.

In conclusion to this section, it should be said that as this field is still emerging, the jargon and definitions are somewhat disparate. Hopefully, the reader will pick up most meanings by the context and normal usage of the commonest terms.

## CLASSES OF CONTROLLER

As explained above, a certain amount of data regarding the controller required for a given job can be gleaned from a simple analysis of the problem to discover the number of I/O lines required. However, there are many controllers in existence with a given number of I/O lines, and it is necessary to consider some other characteristics to narrow the choice down, and ensure that the controller chosen is cost effective and efficient. Some of the criteria involved would include, for instance, size and weight where the product to be automated were a child's toy, or if the controller were to be used within a restricted space. The quantity of internal memory storage may be an important criterion in a data logging situation or where the controller is to contain a large 'look-up' table of values for some of its decisions. An important parameter is the cost of the controller versus the number of units required. Some controllers are cheap to buy if ordered in many thousands, but enormously expensive if ordered singly. Indeed some customised units cannot be ordered in quantities of less than 10 000.

We will look, shortly, at the types of controller which may be chosen, and see some typical parameters which are used to distinguish them. However, before continuing, it is important to consider one of the controversies in the industry, which anyone involved in automation will have come up against at some time or other.

There are two types of controller which often compete with each other in the area of machine and process control. These are the PLC (programmable logic controller) and the ordinary MPU-based system. PLCs are complete, fully enclosed units which try to emulate the ideal black box approach, and tend to be somewhat less flexible than the MPU-based systems. PLCs will be described briefly at the end of this chapter, but general MPU-based systems will form the main part of this book, and are described in detail below.

## SINGLE BOARD CONTROLLERS

This level of controller is chosen for description first as it is the commonest type of controller to be considered in any typical application. Such devices may be too large or too small for some applications, too expensive for others, and so on, but when considering the development of a given application, a single board controller (SBC) will often be the starting point. Either way, consideration of this level of device produces examples of many of the main types of decision which have to be taken in a project. We will look at two real devices, but it should be borne in mind that they only represent examples of types, and should not be considered as the only examples available. In addition, the manufacturers may well change the characteristics of the devices described, and again the description below should not be taken as necessarily correct for the current state of these manufacturers' products.

As its name implies, the SBC is a single printed circuit board with a number of components soldered to it. It will normally require an external power supply unit (PSU) and some means of connecting it up to a larger computer or screen and keyboard in order to program it for the given application. The board will also have a terminal block or edge-connector through which the outside world devices are connected. There is sometimes some interfacing on the board to allow the low voltages of the MPU system itself to be amplified to drive power into external circuits. We will look a little further into the design of the internals of SBCs in Chapter 2 when we discuss controller design in general.

To illustrate the range of the commonest types of SBC, there are plates of the J.P. Designs SBC and the Arcom ARC40 SBC. Both of these companies produce a range of controllers, and the ones chosen illustrate specific characteristics of typical SBCs.

The J.P. Designs controller (Plate 1) is very simple and low in cost. It is not difficult to design a controller to do the same job, but this SBC is so low in cost that it is simply not worth designing one's own for a given application – it makes more sense to purchase such an SBC already made and concentrate on the application itself.

The board contains nine dual in line (DIL) integrated circuits – flat plastic packages with two parallel rows of pins. There are a few other components shown and a double row of pins standing up from the board onto which an external connector can be plugged. This plug takes I/O lines as well as power to the board.

To identify the integrated circuits, look at the two largest chips which are arranged end to end. One of these is the MPU and the other is used to supply 16 I/O lines to the outside world. Memory, for both the controlling program and general data, resides in the three next largest chips arranged side by side in one corner of the board. Each of these chips can contain a

*Plate 1* Single board controller (*courtesy of J.P. Designs Ltd, Cambridge*).

total of 8K bytes of program memory and fixed data, and 2K bytes of variable read/write memory. You should refer to Appendix 1 for the meaning of these terms if they are unfamiliar. We will look at the different types and densities of memory in Chapter 2, as this is a crucial attribute of controllers in general.

The Arcom ARC40 board (Plate 2) is essentially the same in concept, but is an example of a more expensive and sophisticated type of SBC which may need to be considered for an application. It has a maximum of 40 I/O lines, and much more internal memory provision. As can be seen from the photograph, the general view is one of a more densely packed board, and this implies many more functions. There are, again, two large DIL chips, and once again they are MPU and I/O chips. However, in this case the MPU itself is an almost complete microcomputer system with internal memory and some of its pins devoted to providing I/O lines directly. The extra I/O chip then provides 20 more I/O lines to give a very comprehensive I/O structure. There are two connectors for the I/O and power, and the board also contains three extra memory chips giving up to 32K of memory in total.

The main difference between these two SBCs is only fully appreciated by considering the way in which programs are developed for them. In the case of the more economic J.P. Designs board, a separate larger computer system is required on which programs may be written to control the

*Plate 2* ARC40 Single board controller (*courtesy of Arcom Control Systems Ltd, Cambridge*)

function of the board. Such a system is generally termed a 'development system' when used in this way. Thus, a company using this SBC must have access to a full computer system which is configured in exactly the manner that will interface with the SBC. This is no easy matter, as we shall see later. The Arcom ARC40, however, has the advantage of an on-board development system. The MPU chip, as explained above, is a complete microcomputer on a chip – a so called 'single chip microcomputer' – and even includes a full program for allowing the user of the board to write and store programs for running on the board in the well-known and simple computer language of BASIC.

The version of BASIC offered by the ARC40 is of a rather simple nature compared with that found on even the simplest of home micros. However, it is orientated towards control applications, and allows the system to be used to develop highly sophisticated and efficient programs for the use of the board. These programs can also be used on other, cheaper, SBCs in the Arcom range, and as such allow a complete control environment for experimentation, as well as for the development of low cost control applications.

The ARC40 may, from the above, seem to be an ideal solution to any control situation. However, it should come as no surprise to learn that the decision as to the correct controller for a given application is rather more complex than this suggests.

The two boards described above have been introduced to give the reader a feeling for the type of technology which is common in this area of engineering. We will now look at some simple applications of controllers and see where these two controllers might figure.

## Applications for control

Several examples of control applications were mentioned in previous sections, and we shall examine these again in the light of the above information on controllers. One example was that of a large thermoforming machine and another was a washing machine. These are two extremes of the application of controllers to specific manufactured products.

The thermoformer, which is shown in Plate 3, is a machine which takes sheets of plastic, heats them precisely and then forms them around a large heated mould. The model shown is a linear process machine. A stack of flat plastic sheets is placed on the tray shown inside the opening at the end of the machine in the photograph. Each sheet is picked up by vacuum cups and transported into a heating station. There it is heated until soft and then transported to the forming station. Meanwhile, during this indexing, another sheet will have been picked up and transported to the heaters. In this manner, the machine contains several sheets at different stages in the process at any time while production proceeds.

To give some idea of the size of the machine, the viewing windows shown in the side of the machine are too high to be seen directly, and a three foot high platform is normally placed by the machine for viewing. The whole assembly weighs nearly ten tons. There are many pneumatically actuated rams within the machine and some of these can be seen extending from the top of the machine.

The electronic computer control system, designed by the author, had to solve two main control problems. The first was the sequencing of the operations, with its careful timing and treatment of several sheets at different stages simultaneously. This was a fairly complex control problem by any standards and used a specially designed dedicated computer controller to achieve it.

The second problem was that of the heater control. In order to vacuum form the sheet around a mould, some parts of the sheet are stretched considerably more than others. The trick is to impress a high-resolution heat pattern into the sheet so that the highly stretched parts are heated somewhat less than the lightly stretched parts. This ensures that the

*Plate 3* Plastic thermoformer (*courtesy of Shelley Thermoformers International Ltd, Huntingdon*).

thinning over the highly stretched parts is not too great. The correct term is to ensure even 'distribution' of material across the formed product.

The problem was increased by the manufacturer's requirement of impressing the heat pattern into the sheet much faster than normal, by overheating or 'superheating' the material at the start and then reducing the heating temperatures progressively to prevent burning. To further increase the difficulty of the problem, mains variations were to be compensated for, to ensure that the quality of the product did not suffer if, say, one or more of the mains phases dropped momentarily by 10%.

This produced a complex real-time control problem. The two-dimensional heat pattern on two banks, of up to 300 heaters each, had to be calculated, changed on a time scale of fractions of a second, and used to switch hundreds of separately settable mains current heaters. The whole process had to be settable to allow different time-heating profiles to be input from a front panel, and the whole of each heater bank surface had to be constantly displayed to the operator. This is shown, in the Plate 3, as an array of lights on the side of the machine, which in this version did not have as many as 300 heaters per bank.

As the sequencing problem was complex enough on its own, it was decided during the initial systems analysis that a second computer system would need to be dedicated to the heater control. Some further initial calculations showed that even this approach would not yield a full solution to the problem and it became clear that some special hardware would be needed to speed up the control process. The job of this section of the hardware would be to take the calculated heat pattern from the computer's internal memory and realise it as a set of controls to mains TRIAC controllers on the heaters.

An interesting part of the problem was that in addition to the other controls, the temperature of the sheet had to be sensed and used as an overall feedback parameter to the control process. We will see how this temperature was sensed in Chapter 3. The main problem was that a non-contact method of sensing had to be utilised as there was no way to hold a sensor against the centre of a large sagging plastic sheet while it was being heated between two banks of high intensity heaters.

As a final note, it was simply not possible to include the computer control circuitry within the main body of the machine and in general the controller is confined to a large separate enclosure, not shown. This is a common situation in the control of large machinery and will be seen on shop floors throughout manufacturing industry.

This control problem is an example of the larger type of situation where an expensive machine is being produced which often must be customised to the specific needs of a given client. It will have perhaps hundreds of I/O lines for its control and may require large amounts of memory. A certain degree of flexibility is required in the controller used for such applications. The washing machine controller, on the other hand, will generally be reproduced exactly for all examples of a given model, and no flexibility is required in the same sense. It also requires far fewer I/O lines, and little memory.

The main differences between these two examples may appear at first sight to be simply the scale of size of the required electronics. Both SBCs mentioned above will be adequate for the washing machine, and perhaps neither for the more complex thermoformer. This is where an understanding of the technology of the controllers is vital. In fact, it turns out that the washing machine cannot possibly use either SBC in a viable and cost-effective system, while both boards are quite possibly applicable to the thermoformer. In this latter case, two SBCs will be required, plus some extra expansion boards.

The reason behind the decision in this case is based partly upon economics and partly on the technical capabilities of the controllers involved. The washing machine will be produced in vast numbers and each of its components will be pared down to an absolute minimal cost by the production engineers in the washing machine manufacturer. They will want the electronics confined to as small and cheap an area as possible, and preferably only requiring plugging into a standard cable form with no extra expensive assembly procedures. The thermoformer manufacturer, however, while just as interested in cost, has a different order of cost to consider. The assembly of his machinery may take three months per unit, and a few extra hours of assembly makes little difference to him. In exchange, he needs a sophisticated controller which can be programmed and even interfaced differently for each new client.

This brings us to the technology used to produce the controllers. Each of the controllers mentioned above is an unpackaged (sometimes called OEM) board which requires a certain amount of electronic engineering by the user. The washing machine manufacturer, however, could do with a single complete and compact sealed module, containing the minimum of electronics and costing pence rather than tens or hundreds of pounds. This can only be achieved by customising a single-chip computer for his special and invariant needs. We will see how this is approached a little later.

The thermoformer manufacturer is used to dealing with the bare essentials of electrical and electronic systems, and is happy to purchase the SBC level of controller. However, he is not generally an electronic engineer, and he will need to engage a subcontractor to develop the required controller and perhaps even customise the final product to each new specific application. The attendant extra cost is a small fraction of his profit on a sale, and in exchange he is able to provide his machines with a number of special edges over his competition, perhaps reduce his manufacturing costs, and provide an intelligent flexibility to his product range.

In either case, it is the development of the electronics and software for the application which is the greatest part of the controller's cost. The development cost is, of course, directly related to the amount of work which has to be performed, and can be considerably reduced by choosing sympathetic and easy to develop standard equipment such as the SBCs mentioned.

As a final note to this section, it may be surprising to hear that the SBCs can be applied to the thermoformer which has been stated to require perhaps hundreds of I/O lines, or at least somewhat in excess of that offered by even the 40 lines of the ARC40. The reason lies with a method of increasing the apparent number of I/O lines available on a given controller. The technique involved is that of 'multiplexing', which simply means the utilisation of a given line for a number of functions at different times as defined by the MPU system under program control.

An example of multiplexing is in the use of a telephone line to transmit and receive many hundreds or even thousands of telephone calls apparently simultaneously. This is achieved by redefining the use of the line many times per second, and sending parts of the conversation in many minutely separated 'time slices'. The process occurs so fast that the slow reactions of the humans involved, plus some electronic filtering, prevent the continual breaks in their conversations from being perceived. We shall see a somewhat simpler electronic circuit which can expand the I/O capabilities of a given controller *ad infinitum*. However, there is a penalty in terms of speed, but this is rarely a problem in process and product control.

Before considering further levels of controllers, we shall now take an overview of the development of a control application, and see where the problems occur.


## DEVELOPMENT AND OTHER COSTS

As mentioned above, the J.P. Designs controller, though simple to use and low in cost, requires a full computer development system to write programs. The more expensive ARC40, however, contains the basis of an adequate development system on board. An application which can use

| | Qty/yr | Dev. costs | | 1st year unit costs | | Subsequent unit costs | |
|---|---|---|---|---|---|---|---|
| | | Simple cont. | Complex cont. | Simple cont. | Complex cont. | Simple cont. | Complex cont. |
| Product A | 500 | £5000 | <£2500 | £40 | £60-£65 | £30 | £60 |
| Product B | 10 | £3000 | £300 | £350 | £190 | £50 | £160 |

*Table 1.1* Comparative controller costs

either of these controllers would have to be considered on a full cost analysis to decide which is the more economic.

As an example of the considerations involved, see Table 1.1 and imagine a product (A) which is to be manufactured in quantities of 500 per year. If the development were to cost £5000 using a simple controller with a full external development system, this would add £10 to each unit during the first year and, assuming a controller cost of £30 gives a first year unit cost of £40. Subsequent years would yield a unit cost of £30. If, however, a more complex controller were to be used, along with an internal or simpler development system, the unit cost of the controller may be twice as much or more, though the development may be less – say half as much or less. This gives a minimum first year cost of £60 to £65, and £60 thereafter.

As you can see, even if we discount the development costs entirely in the more complex case, it is still the simpler controller, though rather more expensive to develop, which is the more cost effective. Of course, in a case where only ten controllers were to be used in the first year, it would be silly to spend £5000 on development in order to use a cheaper controller if a more expensive controller can be developed for a fraction of the cost.

See Product B in Table 1.1 and imagine that this is a simpler application to develop. Even so, the simpler controller requires a subcontractor with a full development system and a highly trained team of experts to develop the application. The development costs may still be of the order of £3000. The more complex unit, however, with its easier development method may be able to be put together for a tenth of the cost, and may even be able to use fewer technical people within the manufacturer's own staff. Because there is little economy of scale in this case, the controllers will also be more expensive, and this could give a first year cost of £350 for the simple controller, and £190 for the other. Of course, subsequent years will yield a different picture, and it is clear that an initial high investment could be worth while in the long run. However, if the sales figures for the item are to be cautious, the more complex controller allows very little in the way of investment for a comparatively short route to a workable solution.

Naturally, the above costs must not be taken as accurate, but the point should be considered seriously as to how to trade off simplicity and low cost in the controller with higher development costs. In a given situation it may also be necessary to consider the on-going cost of maintenance to both hardware and software, as well as the differences in assembly costs for the final manufactured article. This is certainly so for the thermoformer where the software in particular may need constant maintenance to change it for each new machine.

The considerations above are somewhat orientated around development for a product rather than a one-off for internal process control applications. Similar financial considerations appertain, but in the latter case the more complex unit should be considered on the grounds that it can be developed more easily and quickly, and will be more flexible in the long run. This is not just an intangible benefit – flexibility means allowing the function and purpose of the unit to evolve as experience is inevitably gained in the field of the control of one's own manufacturing processes. Most people in the field would agree that a process control system is not easy to specify fully from the start – a process of evolution is always involved, and expandability and flexibility must be designed in from the beginning.

## DEVELOPMENT TECHNIQUES

To help in assessing the levels of development effort required for a control application, we will now look at some typical development methods.

The two SBCs mentioned above again provide a useful pair of examples of development methods from different ends of the scale. We will suppose that the hardware has already been designed, and the whole target system now only requires a program to burst into life as a washing machine controller, thermoformer, or musical door chime. •

The essence of a program's development is the transfer of a definition of function of the system into logical steps, each of which corresponds to a step which the MPU can perform. The first task, therefore, is to agree a common language between programmer and MPU system. There are many languages available and much controversy as to the best. In the case of the ARC40, it would be useful if the programmer could write in BASIC, as the board itself understands this language.

The next step is to agree a full specification, if possible, between the manufacturer and the programmer. This is one of the most arduous and often underrated stages in the industry. It is also often somewhat inexact in that, as mentioned above, it must contain flexibility. The need for a full initial specification is not just to ensure that the programmer knows what to tell the controller to do. The construction of a specification allows problems in the required function to be considered before work begins, it allows both sides in the project to agree on the work to be performed, it adjudicates later to ensure that any blame for failure is apportioned fairly, and so on. The importance of a specification, and the insecurity which its lack can engender, makes it one of the most important stages in the project. Sadly, this stage is often rushed, ill-defined or missed out altogether and replaced with a vague verbal understanding which may well break down at the first sign of trouble.

The specification, and perhaps some preliminary experimentation, allows the programmer to start building up a set of logical steps for a program. At this point, he or she may wish to try out some parts of modules on the machine, and thus watch the effect of some of the programming which will be used later. This is where the main problem in development comes. The programming ideas are present in the mind of the programmer and it is necessary for these ideas to be communicated to the controller. The most usual method of performing this is to use a keyboard and screen – however, where does one find such instruments lying around, and how can they be connected into the controller? Further, how can these programming ideas be seen to be controlling the target system? It is essential that the programmer can easily see the effect of his programming on the final setup and thus adjust and experiment until the process is complete. It is the job of the hardware engineers on the project to ensure that a simple and efficient system is available for this task. We will now look at how to achieve this with the two levels of controller considered so far.

## DEVELOPMENT SYSTEMS INTERFACE

In the case of a simple application, the ARC40 provides the easiest development path. It only requires a standard form of VDU (visual display unit) with integral keyboard to be connected to some of its pins, along with a standard power supply, and the programmer has immediate access to the internal BASIC interpreter on board. Figure 1.3 shows a typical setup.

Suppose that a set of switches is to be read and a set of lines is to be switched on and off in some target system which is to be controlled. A typical application would be in a simple plastic injection moulding machine. Such machines will probably have externally settable temperature controllers and it is not even necessary for the computer controller to access this information. The control comes down to reading some microswitches and switching some pneumatic controls via solenoid operated valves, perhaps with some inbuilt delays for good measure. This is called a 'sequencer' as it blindly steps through a given sequence of events, controlling each valve as a result of the state of the microswitches.

The ARC40 is simply connected by wires to the switches and interfaces for the valves, and thus can read and operate these devices using the states of its I/O lines. Logic voltages within the ARC40 are 0 V and +5 V, and these are amplified to work the valves. The actual states of the output lines are determined by lines of program code which will form the main control program. To power the electronics, a standard power supply unit (PSU) is used which simply plugs directly into a socket on the printed circuit board (PCB) of the ARC40.

The final link in the chain is to allow the programmer access to the ARC40, and this is achieved by using a standard piece of equipment called an RS232-based VDU. Such devices are very common pieces of equipment, and they can even be purchased on the second-hand market for around a £100. New VDUs cost between three and ten times as much, and may allow many enhancements. However, the method of communication with the ARC40 is the crucial parameter. This is via the well-known and

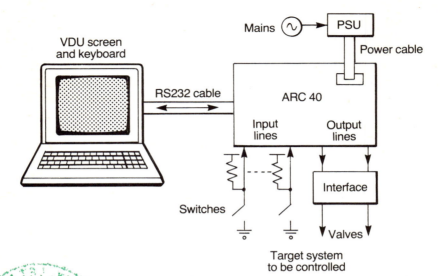

*Fig. 1.3* Controller with internal development system.

standard serial communications method called 'RS232'. This is a means of sending data in binary form, bit by bit, down a single line. RS232 protocol provides sufficient standardisation to cause little problem for the communications.

When a VDU is plugged into the ARC40 in a standard manner, and power is applied, the ARC40 will send a message down to the VDU screen. The programmer is then allowed to type in program steps via the keyboard, which are stored within the ARC40's read/write memory (or RAM – random access memory) automatically. The standardisation to a RS232 VDU removes any worry on the part of the programmer as to how to connect in a screen and keyboard – he can concentrate instead on the program.

The ARC40's internal software allows the program to be edited and run in RAM, at will. This gives the programmer direct and flexible control over the I/O lines connected to the ARC40, enabling him to develop the program in a remarkably short time, which naturally transfers to low development costs.

Once the program has been developed it must be stored permanently in the ARC40. If allowed to remain in RAM, the program would disappear as soon as power is removed from the controller. Thus the program has to be stored onto a different type of electronic memory – such memory is called ROM (read only memory), or in this case EPROM (erasable programmable read only memory). This can be a problematic step, but the ARC40 has special circuitry on board which allows a blank EPROM chip to be inserted into a dedicated IC socket and programmed with the newly developed program. This chip may then be plugged into one of the standard memory sockets on board, and its stored program instantly runs the board when switched on. This converts the general controller into a dedicated sequencer for the given target system.

As you can see from Figure 1.3, the programmer has the ability to view the effects of his program steps immediately – he simply types them in and runs them directly on the target system. This may not always be the case with a controller, and if we now consider the J.P. Designs SBC, the inherent difficulties of developing software for general contollers will become apparent.

Figure 1.4 shows a setup using a controller which has no inherent development capabilities, and no on-board EPROM programmer, as with all low cost controllers. In this case there are several methods of development. The figure shows perhaps the most cumbersome. In this scenario, the controller is interfaced to the target system as before, but it can only accept EPROM-based programs to run it. There is no internal

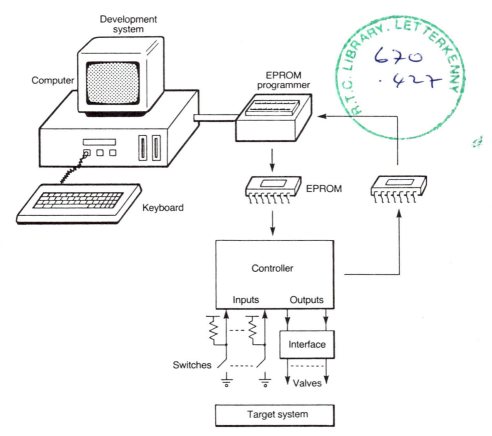

*Fig. 1.4* Controller with external development system.

software within the simple controller to accept program commands from a programmer in real time, and store and run them in RAM. This implies that an external computer, with an EPROM programmer, must be used to substitute for the function, and it is on this machine that the programmer develops and runs his programs.

Unfortunately, in order to test the program on the target, he must first organise the commands to be stored onto EPROM, perform the storage, remove the EPROM from its programmer and plug it into the controller. When the target system inevitably does not perform correctly first time, he is not in direct contact with the process via computer, and thus can do no more than watch the target for tell-tale signs as to the problem. This is time consuming and at the outer edge of efficiency. Indeed, many companies will refuse to deal with a situation which requires this type of approach – they will normally demand a more sophisticated approach before becoming involved. However, this is not to be disregarded entirely, and in the right circumstances, where a very simple application is to be produced, it involves no extra hardware and interfacing over and above the somewhat costly main development system itself.

Another approach is shown in Figure 1.5. Here, the approach is to dispense with the controller itself entirely at the start. The development system has to be amenable to being wired physically to the switches and valve interfaces of the outside world, in exactly the same way as the controller itself. This effectively provides the development system with an 'emulation' of the target system. It allows the programmer to control the outside world directly again, and allows him to check each stage quickly and efficiently and even insert tell-tale modules into the software to use the

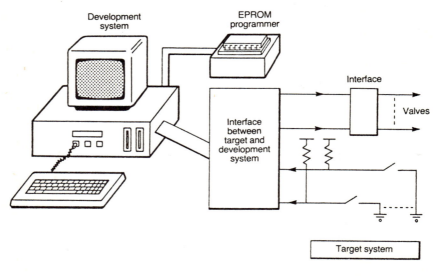

*Fig. 1.5* Interface between development system and target.

development system to debug the software as it is running.

From the programmer's point of view this is similar to the situation of Figure 1.3. The major difference in use is that the software contained on a large and sophisticated development system, as opposed to the internal and restricted memory of the controller itself, will provide many more facilities and aids to the programmer. This can be crucial in developing the larger types of control application.

There is, however, a problem inherent within this system which makes it somewhat less preferable to the next method to be described. The problem comes from the simple fact that the actual final controller itself is not being used during development. This means that no matter how efficiently the software may work in this emulation of the real thing, there is no guarantee that it will exactly mirror the situation when converted into an EPROM, using a programmer, and plugged into the final controller itself. There are many technical reasons why this is a serious problem, and though it may appear to be straightforward to ensure that the development system is wired and configured exactly as for the final controller, it rarely works out so easily. The problems are sometimes so intense as to make the entire effort a marginal improvement over the last method.

The final method of development, which solves many of the most serious of the above problems, is described next. The trick here rests with the observation that the main problem in the above development methods is to be able to transfer memory contents produced by the programmer into the memory of the controller. In the case of Figure 1.3, the internal software of the controller provided the means whereby the programmer typed directly into the internal memory of the controller, using the controller as the development system. The next method achieved the memory transfer by hand – via EPROM chips. The third method above tricked the outside world into believing that it was being controlled by the controller, but in reality the memory contents, containing the program, still resided in the development system until again transferred by hand via EPROM.

Figure 1.6 shows how a previously arranged common memory area can be used to allow the programmer to develop the program in normal development system memory, and then switch that same area of memory over to the controller to run the program on the target system. This has many advantages, but there is still one important facility which may be missing – that of using the development system in real time, i.e. as the program is running, to diagnose problems which may occur. While the program is running on the controller, it will be switched out from the development system. However, with the sophistication of the development system constantly at hand, the scope for being as near to the ideal as possible is great enough to make this method fast and efficient and highly recommended. It also has, once again, the advantage over a controller's own development system that there is a large and sophisticated system

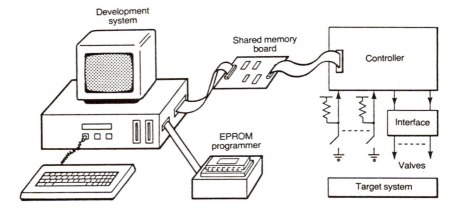

*Fig. 1.6* Shared memory in development.

being used for the development which always increases a progammer's efficiency considerably.

The main problem with this last method is finding a suitable interface board for the memory sharing hardware, and finding a method of connecting it into a development system. Most personal computers, for instance, will not react favourably to their internal memory being shared over perhaps dozens of electronic wires with an external piece of electronics. However, it is one of the best ways to develop a simple controller, and should be considered seriously where possible. Furthermore, some controller manufacturers produce the necessary memory sharing hardware for their controllers, and J.P. Designs is one of these. Of course, the problem may still remain of finding a suitable method of plugging it into a given development system, but, again, the controller manufacturer should be able to advise on this point.

This completes the overall look at methods of developing control applications. Some variation on the above categories of development method will almost always be found in a given project, and a certain amount of care in choosing hardware and software which develops easily is worth while from the point of view of ultimate cost.

## OTHER LEVELS OF CONTROLLER

The above types of controller, the SBCs, were chosen for an introduction to the field of control because they are widely used, and may well be the most cost-effective method of starting a control project even though ultimately destined for another approach.

We will look at some other types of controller in this chapter and the next, in order to give the reader a general feeling for current types and some help in assessing the correct level for a given task.

The first level to be considered below is a development on the SBC, and is simply a compacted version of the SBC approach. It has been mentioned before: namely the single chip controller. We will then examine more complex boards and, finally, we will examine PLCs. In the next chapter we will look at multi-board controllers.

Another level of controller not mentioned so far is the so-called 'discrete level' of control circuitry. This involves non-MPU electronic logic components, and though of some use in certain circumstances, we will concentrate on the MPU-based solutions in this book.

## SINGLE CHIP CONTROLLERS (SCCs)

We are considering control applications in this book, and hence the term 'single chip controller' will be used instead of single chip computer. They are, however, one and the same, and we will use the convenient abbreviation SCC for this type of device.

We cannot look at the internal construction of SCCs, as not enough of the internal make-up of the black box which constitutes a controller has yet been described. However, we can look at their characteristics and economics and discuss their development.

There is little point in showing a picture of an SCC as it will simply look like any other IC – i.e., commonly a flat black plastic DIL package with pins on two sides. There are many varieties of SCC, and some have been in service since near the inception of the MPU itself. When the MPU was produced, it was not long before the industry realised that the ultimate challenge, at that time, was to integrate all the external devices, such as memory chips, I/O chips, clock components, etc., onto the same piece of silicon. Since then, the scale of integration has steadily increased, and some considerably more complex chips have been produced.

The main area of use of such devices is still in the area of control. They are normally a little restricted by the imposed density of function, and there is little advantage, for instance, in attempting to further integrate the chips of a modern personal computer. However, with the advent of the market for the physically smaller type of computer, this may be one of the major growth areas of the next few years.

We will be concerned with chips which essentially compact a complete single board controller onto a single chip. As we are not yet concerned with the internal electronics of MPU systems, we will just look at some examples and discuss their particular attributes.

The first device to examine is the Zilog Z8. This is chosen purely because it happens to be the basis of the Arcom boards above, and as such can be developed using an ARC40.

### Zilog Z8, and SCCs in general

Zilog is one of the best known manufacturers of MPUs, having produced one of the most powerful and widely used small MPU chips in the '70s – the famous Z80 microprocessor. You will see this, even now, figuring in the internal architecture of many small personal computers and other devices. The Z80 was designed to use an enhanced version of the industry standard Intel 8080 instruction language, and as such had an immediate application in the market as it could run 8080 programs directly.

As a departure from the Z80 and its family of associated devices, Zilog designed and manufactured several other MPU products, of which the control orientated Z8 is one. As with all the single chip microcomputers, there are many different versions of the basic device, containing different types and amounts of memory in particular.

The single chip idea is to concentrate onto one chip all the peripheral circuits which would normally be needed to be connected together along with the MPU to produce a microcomputer. This is a straightforward idea, and its implementation is only dependent upon the practical limits of the density of IC production.

The main parts of a microcomputer comprise MPU, clock circuitry, memory, I/O and some extra logic we will examine later. The clock circuitry is simply to produce an oscillator which controls the timing of each cycle of the MPU's activity. There will normally have to be two types of memory included – RAM and ROM. Thus a minimum system will have MPU, clock, RAM, ROM and I/O. This can be achieved with around four chips and a few extra components in a simple system. Thus an SCC must have at least this collection of circuits on its chip to make it worth while.

The Z8, as with most single chip computers, includes all of these facilities, and it is this fact which makes the use of SCCs quite different from that of normal MPUs. The reason is not simply that fewer physical connections are needed when an SCC is used as the basis of a controller, but the fact that ROM is integrated (or 'masked' – see Chapter 2) onto the chip along with the other facilities. This poses the most serious problem confronting the SCC system designer. The ROM must be present to contain the controlling program, but the contents of this ROM cannot simply be programmed onto the SCC as with an EPROM. The contents of the ROM section normally have to be included on the chip during its manufacture.

Naturally, as with most mass production methods, the economics of the process involve the ordering of many thousands of the devices before a chip manufacturer will commit to production. The semiconductor industry is not able to produce small quantities of ICs economically, and this means that the development cycle has to yield a program which is no less than 100% correct before the part is ordered. Any mistakes in the program will be reproduced exactly and unalterably onto thousands of chips.

The main question is, why do manufacturers not simply integrate EPROM onto their SCCs, and save all these problems? The answer is simple – they do, but an EPROM cell uses considerably more chip area than a simple 'masked' ROM cell, and the resulting chips are either highly restricted as to their use, or very expensive. Also, each of the thousands of chips would have to be programmed – an expensive task. In the case of a low cost low volume controller, it is cheaper to use several very low cost chips soldered to a PCB and achieve the full facilities of an MPU system than to design with an expensive and restricted SCC with internal EPROM.

The point is that an SCC should be considered for ultra low cost high volume applications where high initial costs will amortise out to low premiums in manufacture. Having said all this, however, there is an area of overlap which is catered for by the chip manufacturers. For instance, as mentioned above, there are usually several versions of the SCC – there will usually be a version without ROM. In addition, SCCs have a restricted internal RAM, because the RAM cell requires a large chip area. Thus, a middle road to take in controller design is to choose a version of an SCC which has no internal ROM, thus making the chip cheap in small quantities. External RAM and ROM can then be added, to produce a cheap expandable system with a minimum of external connections. If the application subsequently goes to high volume, compatibility across the range of different versions of the SCC allows the program to be adapted, with ease, to being masked onto the SCC for an ultra low unit cost product.

Of course, there are some control applications which will not be able to

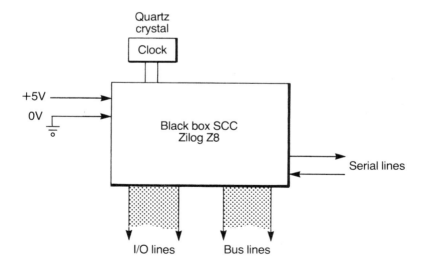

*Fig. 1.7* Z8 functions.

work with the few I/O lines of an SCC, and it is then necessary to look at a fuller MPU system with more external circuits.

Leaving aside the memory capabilities, Figure 1.7 shows the main external lines which the Z8 supplies in all of its versions. At the top left hand side of the box there are electrical lines for power supply, which must be a stabilised +5 V supply. There are also two lines to be connected directly to a quartz crystal from which internal circuitry generates the required oscillating clock. The right hand side of the box contains some I/O lines called 'serial lines' which can be made to communicate directly with a VDU, as discussed above, for external program development, for instance. The bottom of the box shows the most important lines in the system – the main controlling I/O lines. The output lines can be set to +5 V (logic 1) or 0 V (logic 0) under program control and hence be amplified to control any type of outside world device. The input lines can read these voltage levels (logic 1s and 0s) to allow the state of a switch or sensor to be read into the controlling program for feedback from the system under control.

The final section shown in the figure is that of the bus lines. These lines are described more fully in Chapter 2 when we look at the internals of a controller, but for now they should be regarded as a means of expanding the basic controller to include further electronic devices such as extra memory and I/O.

The essential point is that it is only necessary for a controlling program to be written to cause the controller to shuffle 1s and 0s around on the I/O lines, and the Z8 becomes any desired type of controller. As we shall see, such control is not restricted purely to discrete logic levels, with switches and on/off devices; analogue (continuously varying) voltage levels can also be controlled and read at will.

The black box approach here does not show the memory facilities, and we will look at these now. The Z8 always has 128 bytes of RAM memory included on the chip, and much skill is often required in memory use to make this amount suffice in a control situation. If more is needed, then a purely single chip level of controller will probably not be possible. However, many types of simple control can make do with this amount of RAM, which is often used purely as in a programmable calculator – for storing intermediate values during a given process.

The provision of ROM is a more complex matter. One version of the controller allows a total of 2K bytes of ROM to be masked onto the chip, and this is often quite enough for simple applications. Another version allows an EPROM to plug into the top of the chip's surface in a 'piggy-back' configuration. The controlling program is produced using one of the development methods described above, programmed onto a standard EPROM, and this is then plugged into the chip itself for testing. When the program is considered correct, it is sent off to the chip manufacturer who then masks it onto the chips supplied to the purchaser.

This ensures that 100% testing is a possibility.

Yet another Z8 version (as used in the ARC40) comes with masked ROM containing a control program allowing the user to write a program for the chip in a simple form of BASIC. There are only around fifteen command words available, but these are sufficient to produce remarkably sophisticated control programs, particularly in that they can link in 'machine code' routines written elsewhere. Machine code is the internal language of the MPU, and much of the MPU's function concerns fetching such instructions from a suitable memory location and executing the implied commands (see Chapter 2).

Arcom have enhanced the facilities of the BASIC version of the Z8 somewhat by providing their own EPROM-based control program (or 'monitor program') and the result is a highly sophisticated programming environment using a VDU.

## Self contained SCCs

The Z8 above is an example of an expandable SCC which can be formed into a larger multichip controller, or used in a minimal single chip system. The original idea of the SCC, however, was somewhat different. The early SCC approach was to produce a chip which was as nearly self contained as possible, not expandable, but capable of cheap high volume use in such low cost systems as musical door chimes and children's toys.

There are many examples of such devices and they form a most important section of the market for control devices. However, they are considerably less accessible in terms of development than the Z8, and similar devices. One of the first musical door chimes, for instance, used a controller called the TMS1000. This chip was produced by Texas

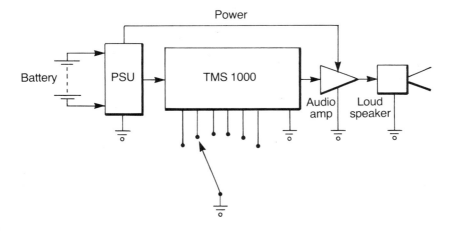

*Fig. 1.8* Musical door chime.

Instruments, and subsequently used by them in many products, including their famous 'Speak and Spell' speech synthesis toy. The TMS1000 contains MPU, RAM, ROM, clock and peripheral circuitry and I/O to form a complete system. It cannot generally be expanded to include further memory and I/O by using 'buses', and it is intended to form the basis of a complete and ultra low cost system.

As an example of the type of design which would be needed for an application on this level, Figure 1.8 shows how simple a circuit can be using this device. This is not meant to be an actual circuit, but it does demonstrate the level of simplicity of this application. The TMS1000 has several I/O lines for general use and these are connected directly to a rotary switch which simply sends 0 V to one of the lines depending upon which tune is to be selected. Another I/O line connects to the input of an audio amplifier, a single IC, which then connects to a loudpseaker. The PSU is simply a few cheap components to stabilise a dry battery which is contained within the unit's housing. The result is a single cheap PCB which contains everything for the product.

The internals of the TMS1000 contain all the complexity of design for the project. The programmer would have to be supplied with a full emulation system. No work can be done on the TMS1000 itself as it is not capable of being connected into a development system. The chip manufacturer has to supply a purpose built development system, or agree to develop the program for the project. The programmer uses this development system to produce a program which simply reads the I/O lines to determine which tune is to be played and then sends out to the audio amplifier pulses which, to human ears, sound like the tune selected. Such programming is straightforward and development is quick but quite expensive due to the cost of a development system.

This example, however, points to an entire class of control problems where a self contained chip will do all the work and a few external components complete the product. A washing machine controller would be amenable to this approach, or that of some parts of an automobile system. Any application which is simple and to be produced in high volume will conform to this approach.

Another common device is produced by General Instruments (GI) and is called the COPS system. Again, the chip is very simple and cheap in high volumes, but must be developed using GI's purpose built development system. Thus users tend to employ GI themselves for the development work, rather than attempting to produce the product themselves. In each case of this type of controller the development cost is high, and only high volume applications are considered worth while by the chip manufacturers. Given this situation, however, the unit price per controller, even with development, can be very low.

## Another SCC

Most major MPU manufacturers make some type of SCC, and the devices described in this book are not meant to form a complete list. Many of these devices are similar, and the aim here is to present the reader with enough information to be able to decide on the level of controller which would best fit any given application. Having selected a particular level of controller, it is important to spend some time looking at the different manufacturers which exist and assessing their products a little more deeply.

We will now look at one more widespread SCC – the 8048 family of devices. This range was originally manufactured by Intel, but has subsequently been licensed out to other 'second source' manufacturers.

Figure 1.9 shows a black box sketch of the 8048 family. The family is being changed continuously, but the two main devices are the 8048 itself, and the 8035. The 8048 has masked ROM, and the 8035 has not. Each chip has a block of 64 bytes of RAM, two I/O blocks of 8 bits each, 8 more bus lines which allow expansion or may be used as I/O lines, and a set of control lines for expansion. In addition, the 8048 has 1K bytes of masked ROM space. Figure 1.9 shows once again how the clock is formed on the chip and just requires an external quartz crystal (labelled XTAL) for completion.

The development of projects using this device is not as easy as that for the Z8, as there are no standard development products available apart from large and expensive systems from the supplier. In addition, as you can see, the 8048 devices have fewer facilities than the Z8. However, this family is important for two reasons. Firstly, it is so widespread that if you examine the chip at the heart of many of the everyday items around you,

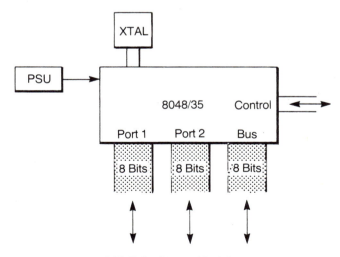

*Fig. 1.9* 8048 family as a black box.

there is a good chance that you will see the number '8048' on its upper surface. Secondly, it provides a good example of the breadth possible in a family of controllers.

Due to the policy of continuous change prevalent among most semiconductor manufacturers, some of the following version may no longer exist, and there may well be others which have been substituted. However, the following shows a roundup of the various versions which were available at some time.

The basic 8048 and 8035 chips exist in several technologies, for different applications. There are the 80C48/35 versions which are low in power consumption for battery applications. In the standby mode they consume a few hundred microamps and need around one tenth of the current of the normal version during use. In addition to these versions, there are the 'automotive' versions called the P80A48L and the P80A48H/P80A35HL, among others. These versions have different maximum clock frequencies and are especially rugged for automotive use. There are normally also other versions of popular chips which are designed for military applications.

In addition to these generally pin-compatible 8048 and 8035 chips, there are other versions in the range with more or less characteristics. For instance, the 8020 is a physically smaller (20-pin package) version of the 8048, with fewer machine code instructions; it is cheaper and generally designed for the more cost-conscious consumer market, such as washing machines. The 8021 version of the family has a cut down machine code set and some special I/O lines and internal counter timers which make it useful for controlling mains devices. The 8022 version is designed for more complex applications. It has special I/O lines for reading analogue (continuously varying) signals, as well as more memory on chip.

Another important version, the 8748, contains EPROM on chip and thus allows the control program to be tested on the final system directly before committing to manufacture.

Other ranges of control product also exist in this proliferation of devices. There are the larger memory products of the 8049/8039 chips, the greater sophistication of the 8031/8051/8751 versions, and so on. The extent of this range of control products is prodigious and typical of a company which pioneered, in many ways, the MPUs and silicon chips upon which all our current concepts of these devices are founded.

## LARGER SINGLE BOARD CONTROLLERS

In addition to the many ranges of smaller control boards and chips which are described above, many companies are now producing single boards which have the power of some of the personal computers. These boards require full development systems to take advantage of them, and are generally used in the upper end of the control spectrum. They normally cost a minimum of two to three times as much as the smaller ARC40 level of board. Their greater facilities allow their use in such applications as complex data logging, calculation and analysis applications, or in the processing of large amounts of information at high speed, for instance in complex communications projects.

An application in the process control world which would be amenable to this larger type of board would be the control of the heaters in the upper end of the market thermoforming machines mentioned before. The problem is to control many hundreds of separate fast acting quartz element heaters, in fractions of a second, according to a complex two-dimensional pattern of temperatures. The pattern of temperatures created on a piece of flat plastic allows it to stretch correctly over the surface of the mould. Too much heat at one spot may make the plastic thin at that place, and too little will cause it to be too thick. The computer's function is further complicated by the need for a fast acting time-dependent method of 'injecting' the heat into the surface by careful and controlled overheating at the start of the process.

The main point about this system is that there is a fair amount of continuous calculation and feedback from the process to be performed, simultaneously with the control of the heater elements themselves. A fast and efficient MPU will find this type of task easier than the smaller, slower MPUs. In fact, this type of task is probably better performed by the 16-bit MPUs on the market, with their more complex machine code sets, and faster operation.

A typical board level product for this application is one of the later boards from Arcom or J.P. Designs. Such boards are based upon MPUs such as the 68000 and Z8000. These are 16-bit processors, as opposed to the 8-bit processors used in the smaller boards above. We will see the difference between these two types in Chapter 2, but for now it is sufficient to know that 16-bit processors are generally more powerful and faster than 8-bit processors.

Many controller manufacturers have a range of controllers from 8- to 16-bit devices. J.P. Designs has a couple of such controllers, for instance. One of these uses a 68000 MPU and provides up to 32K of RAM, up to 96K of EPROM, 24 lines of I/O and many more features to allow quite complex applications to be covered. They also produce a Z8000 based controller which has even more I/O and RAM. Arcom also produce a complete range of 16-bit controllers, based around the Z8000. As with many such manufacturers, there is also a further range of peripheral boards for various applications.

However, it is important not to use a system which is too complex for the application. This simply wastes money and effort and may make the final product too expensive to be used. Great care should be taken to pare a project to its basics before starting design.


## PLCs

PLCs are designed to be as near the ultimate black box as possible in control. The user does not need to understand any microelectronics to use them. A PLC can be programmed to switch a set of output lines on and off depending upon internal adjustable delays and the states of a set of input lines from the outside world. To simplify further the problem of use, many PLCs simulate relay logic and allow the program to be written in classical 'ladder logic'. Ladder diagrams were used to design the logical connection of relays when controllers were designed using these devices as the basic gates of the system. These have now largely been replaced, either with simple small scale integrated circuits or with complete micro systems.

In a very simple sequencing operation, it may be that the research and development for a PLC solution will be lower in cost than that of an MPU system. However, it is not unfair to say that many PLC systems are in place because the managers and engineers involved in using them could understand them considerably more easily than even the simplest MPU system. Some large machines have standard PLCs built into them as modules bought from the PLC manufacturers. They effectively provide complete packaged solutions to the control problems, with no need for the user to understand more than their external function.

PLCs come in two main complexity levels. The lowest level is a box containing a sequencer whose operation is decided, perhaps, by a patch board on the front of the box. Some screw terminals are provided for connection of the external devices to be controlled and the system is simply plugged in, programmed by placing a few pins in the patch board, and away it goes.

The more complex systems will supply a keyboard and display, or even a complete VDU system with keyboard and screen to allow the user to program them fully. Such systems provide modules to read and control high power equipment, read and produce continuously variable voltages, and perhaps interface to special sensors. Variable delays may be programmed in, full feedback loops with the controlled machinery are possible, and the whole program may even be retained by internal batteries for a long period to ensure that the system continues working correctly, but may be altered at any time.

The effect of using a PLC is often a highly cosmetic and satisfying solution which is simple to program and solves straightforward sequencing problems with little fuss. However, the sophisticated and flexible solutions which an MPU-based system will supply, along with low cost in comparison, are not available using the PLC approach. PLCs were designed to bring sequential relay-type automation into the world of microelectronics with the minimum of work. The programming and operation are designed to allow a relay-based system designer to transfer directly to electronics without having to learn a whole new sphere of engineering. However, the costs in terms of flexibility as well as financially are high.

This brief review has been included to supply some of the information required to allow a comparison to be made with the MPU approach, though ironically almost all PLCs are MPU based. For further information on PLCs, you should consult the indicated companies in Appendix 2.

## Chapter Two
# Controller Architecture

### INTRODUCTION

As previously explained, this book aims to introduce enough background data and general control information for a mechanical or electrical engineer in a manufacturing company to choose the correct direction of attack, and level of controller, for a given project. It also aims to awaken ideas for the use of high technology in areas which may not have been apparent previously.

This chapter provides a down to earth introduction to the design and construction of controllers and peripherals of various levels. It aims to prepare the reader for understanding how modern 'intelligent' devices are put together in general, but without the detailed electronic descriptions which can be found in one of the specialist books in the Bibliography. The information included should allow the reader to understand the literature on controllers, and thus aid in the task of choosing the correct level of hardware to apply in any given circumstances.

The information contained here is largely aimed at the small to medium sized project, and generally describes the internal architecture of MPU-based devices. The last part of the chapter uses this information to show how a control system is designed from scratch.

Background knowledge of this form is of particular assistance in interfacing with the designers during a project, and aids considerably in making the decision as to the correct devices to use for a given problem. However, the descriptions are fairly technical so as to ensure that they are not superficial, and if the internal construction of the controller itself is not of immediate use to the reader, the first part of this chapter may be skipped and used as reference as required. However, the case study described towards the end of the chapter should be valuable to anyone considering a control project.

Appendix 1 provides an introduction to bits, bytes and binary numbers for those to whom these concepts are unfamiliar.

## THE COMPONENTS OF A SYSTEM

Figure 2.1 shows a block diagram of an MPU-based system which will be familiar to any designer of such devices. It contains the main common components of any system, whether a large multi-board computer or a single chip controller. The difference between these levels is simply the compaction, or integration, of the various parts shown.

The main parts of the system are seven blocks as shown and a 'bus' system. The seven blocks include MPU and clock, two types of memory, I/O and peripheral functions and a block labelled 'address decoding'. The bus system consists broadly of three separate groups of electrical lines. These are 'address bus', 'data bus' and 'control bus'. These lines are used to interconnect the blocks which comprise the system.

### Memory and the MPU

The heart of any programmable system is the MPU itself. The function of an MPU is to fetch and execute program commands and electronically control the entire system. The MPU thus has two main functions – software execution and hardware control. In fact, it is this ability of the MPU to 'control' which leads to the full gamut of control applications which are mentioned throughout this book.

One of the main reasons for the MPU's dual function (software/-

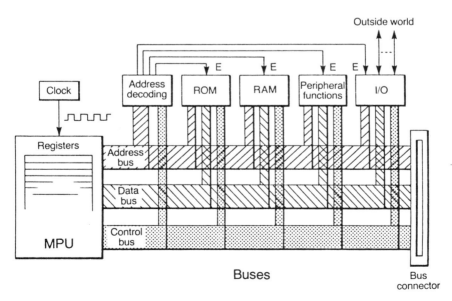

*Fig. 2.1* MPU-based system block diagram.

hardware) is that the software (a set of program commands) is inevitably contained in the hardware (a block of memory) external to the MPU itself. In general, the MPU has to find a method of copying the commands, one by one, from the memory chips into its own internal memory space. Having done this, it can interpret and act upon these commands.

The internal MPU memory space is in the form of a set of 'registers', as shown in Figure 2.1. All the commands which the MPU executes have some effect on the MPU's registers, and as such these memory locations have a special position within the system. In fact, it is only through these registers that the programmer has any access to the MPU. As far as he is concerned, they can be considered as containing the complete current state of the MPU at any instant.

The memory blocks (RAM and ROM) shown in Figure 2.1 are on a different footing from those of the MPU registers. It is necessary for the system designer to find some method of physically integrating these memory blocks into the system electronically, while the MPU registers are automatically included. This is achieved by using the concept of an electronic label.

Each memory location within a memory block is labelled electronically by a code called an 'address'. Electronically, this is simply a set of 1s and 0s applied to some special pins of the memory chips called 'address bus' pins. Each program instruction, or piece of data, within a memory chip has a unique label and can be addressed by placing the correct binary pattern on the address bus. We will see how this works, and how the data is read, by considering the MPU activity of fetching and executing instructions contained in memory.

### The buses and the fetch and execute cycle

The activity of fetching a program command from memory to the MPU registers is a purely electronic problem, and has little to do with the programmer. In fact, while a programmer is developing his program to run a controller, he is supremely indifferent to these problems of the internal electronics of the computer system – he assumes that his program commands will be fetched into the MPU and executed automatically. Only the system designer has to worry about how this occurs. He is responsible for interconnecting the integrated circuits of the controller correctly to allow the MPU to perform its tasks.

As we have seen, EPROM chips can contain these program commands, and Figure 2.1 shows a block of ROM connected into the system buses. We use the term 'ROM' as a generic term, here, to mean any type of read only memory, of which EPROM is an example. As you can see, there are several other blocks in the diagram, each connected into the buses. It is important to realise that all the blocks are physically connected together by these buses, which are just conductive electrical wires of one form or another. Thus, if you use a continuity meter and connect it to, say, the third line down in the data bus, you will find that this bus line connects through to the third data bus pin on each of the blocks shown connected to the data bus, including the MPU. Naturally, the third data bus line is insulated from the fourth, the second and all the others, or complete electronic anarchy would prevail!

The use of buses to interconnect different blocks within a system is one of the greatest leaps made by modern electronics. It has the advantage of providing an almost mechanical method of grouping like functions together, as well as allowing a certain standardisation on interconnection between wildly differing components. Busing also simplifies the design, understanding and troubleshooting of any system. There is one problem, however, which must be solved in any bus system – how to stop two or more blocks from trying to output data to the bus lines at the same time, and causing electronic interference and confusion.

As we shall see, this problem is solved by the use of 'tri-state' devices, and 'enable' pins, with the instant by instant organisation of the whole system down to the MPU. This organisation actually occurs on a submicrosecond basis in most modern MPU systems. The trick is to give one block, the MPU, the task of deciding exactly what each of the other blocks in the system is to be doing at any given instant. Suppose the MPU is running a program, and it has just completed the current program instruction. Its next task, and one of the most common which it performs, is to fetch the next instruction from the ROM and execute it. This is the fetch and execute cycle.

Figure 2.2 shows a common type of diagram, used in the literature, for describing the activities of various parts of a control system – it is a 'timing diagram'. The timing diagram shown is that of a typical 'fetch and execute

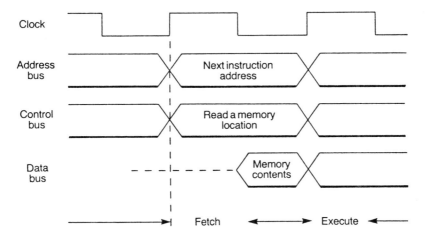

*Fig. 2.2* Fetch and execute cycle of an MPU.

cycle'. The top line of such diagrams is almost always the clock of the MPU system against which all actions are referenced. The exact number of clock cycles required for each of the activities described below varies considerably from MPU to MPU, but the diagram shows a possible pattern.

When the MPU wishes to fetch its next instruction, it usually knows where it will be, because it has just fetched and executed an instruction from the previous memory location in the program. Thus the MPU places the binary pattern for the address of the next instruction on the address bus, as shown. At the same time, it uses its control bus, another set of lines, to alert all the blocks in the system that it is about to read in the contents of a memory location. These two actions affect the address decoding block, shown in Figure 2.1, which sends a signal out to the ROM block where the required address resides. This signal is sent to the 'E', or enable, pin of the ROM, and causes only that block to wake up and accept the address bus pattern. The ROM chip then sorts out which memory location is being contacted and sends the contents of the location back to the MPU along the data bus, as shown. This completes the fetching of the command, and the MPU stores it in a register, and then executes it.

The timing diagram summarises this set of events, and its sequentiality is determined by the upward and downward transitions, or edges, of the clock waveform.

As mentioned above, all the blocks are connected to the data bus, but only one block (ROM) was activated by the address decoding under MPU control. The other blocks were in an inactive state called 'tri-state'. This effectively leaves them electronically unplugged from the system and they cannot interfere. Only when the 'wake-up' or enable signal is received on their 'E' pins can they come out of this state and become active. The word 'tri-state' refers to the ability of a given logic line to take on one of three states – 1, 0 and 'unplugged'.

All devices which can output to a data bus will have this tri-state ability, and it is up to the MPU and address decoding block between them to ensure that only one block has its 'E' pin activated at any given time.

As you can see, the data bus is used here to allow the program instruction to pass between the memory and the MPU. If the instruction needs further data from memory or from the I/O lines, a further fetch is performed from the relevant block, again via the data bus. In the case of an instruction which requires the MPU to place some data in memory, or adjust the lines of an I/O device, the data is then sent from the MPU along the data bus to the relevant block in the same way. The MPU thus uses the data bus as a bidirectional data highway. The MPU uses one control bus line to say whether the data is to be read into the MPU along the data bus, or written out from the MPU. This line is called 'R/W' or the 'read/write' line. Again, the MPU is fully in charge at all times.

As with all the buses, the data bus is a set of parallel electrical lines. The width of this bus determines how much data can be fetched from memory at a single time. Thus a data bus having sixteen lines, for instance, can fetch sixteen bits of data in a single cycle. A data bus of only eight lines can only fetch half as much data. This is important in determining the maximum speed of processing of an MPU system. Thus we talk of '8-bit' or '16-bit' MPUs. The majority of simple controllers, including the first two described in the last chapter, are based around an 8-bit MPU. The more complex controllers are based around a 16-bit MPU, and it is the width of the data bus which is being referred to.

As can be seen, the timing of the MPU's activities is set by the frequency of the oscillating clock, each transition of the clock signals the next stage in the MPU's cycle. Thus the speed of the clock determines how fast the MPU acts. If the clock is slowed down, for instance, the MPU does not 'notice' any change − it simply performs its activities at a lower rate accordingly. In fact, many MPUs have the ability to be 'frozen' simply by stopping the clock at the correct time. When the clock restarts, the MPU carries on as if nothing had happened. Some MPUs, however, will forget their memory contents if the clock is slowed down too far or stopped.

The MPU clock's electronic complexity is determined by two factors. Firstly the MPU itself, and secondly how important stability of timing is to the process under control. For instance, some MPUs can make do with a few cheap passive components for a clock, and internal circuitry generates the oscillations. This produces a clock, the accuracy and stability of which are not assured. Thus if critical timing considerations are required, this is not adequate. However, most MPUs will accept a quartz crystal and this is used, along with circuitry on the chip, to generate an accurate clock.

Another method of producing an accurate and controllable clock is simply to use a complete external oscillator circuit, and feed the oscillations to the MPU directly. This allows the clock to be slowed or stopped if necessary. Most MPUs are quite versatile, and they will either accept just a quartz crystal, or input from an exernal clock.

## Memory

We have already seen some examples of memory and its application. We will now look at this crucial element in a little more detail.

It was the nineteenth century's breakthrough of the 'stored program computer' which heralded the full gamut of today's plethora of computing devices. Babbage's concept was simply that a programmable machine should store internally a list of instructions, and then proceed to execute them from start to finish. By storing the instructions within the machine, these instructions could be manipulated and changed at will, and then left to control the machine. As we shall see later when we examine software hierarchy, the machine is then capable of running a program which itself is used to aid in the construction and storage of further programs for the required applications. In this day and age, we are used to complex and highly accessible computers which allow a programmer to develop intricate software designs using the machine to aid him in the task. This is the present reality of the stored program computer.

Central to the idea, therefore, of the modern computer is memory. Indeed, in many important everday uses, the computer is nothing more than a huge and intelligent store for information. These are the so-called 'data processing' applications. There are areas of control which also need a large memory capacity, such as data logging and storage of large tables of information for the control of a given process. In contrast, many small product control applications in particular require very little memory for data storage. However, the common feature of any computing system is the program memory.

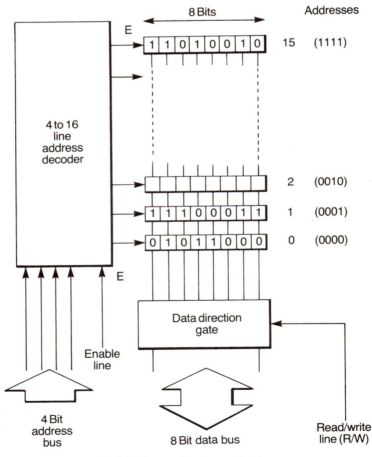

*Fig. 2.3* Memory block organisation.

There are two main types of internal computer memory – read/write memory and read only memory. As the names imply, the first type allows the contents of the memory to be changed and read at will while the second type cannot be changed – it can only be read. Read only memory is generally called ROM, but, confusingly, read/write memory is called RAM (random access memory) for historical reasons. An important difference between these two forms of memory is the fact that RAM generally loses its memory when power is removed, while ROM retains it without change. These attributes are a consequence of the current technology used to produce these two forms of memory device. In fact, it is now possible to produce RAM which does not lose its memory when power is removed, but the technology of such devices is not yet sufficiently advanced to allow them to be used generally. The main problems are expense of production, density of memory and memory speed. This last is important for determining the maximum frequency possible for the MPU's clock in a system. If a high speed MPU clock is used with slow memory, the fetch and execute cycle can 'beat' the memory and the whole system ceases to function.

To understand a number of concepts, which will also be useful in looking at I/O later, we will examine a simple sixteen-location memory block (see Figure 2.3). This block has sixteen locations, each of which is capable of storing eight bits of data. Each bit, of course, being purely a 1 or 0. The locations are shown as a vertical stack, each labelled with an address which is shown both in ordinary decimal notation (0 to 15) and in binary notation. The importance of the binary notation is that the four binary digits of a given address are exactly the electrical pattern of 1s and 0s which appear on the address bus to contact that location.

The address bus shown has four bits because there are just sixteen different binary patterns which can be placed on four lines, and these sixteen patterns are used as the addresses of the memory locations shown. Thus, for instance, the third location (numbered with a 2) has binary address 0010. To contact that address, the sequence of events is as follows. The MPU has to arrange for 0010 to be placed on the address lines shown. These would normally simply be connected to the lowest four lines of the MPU's address bus. At the same time, the MPU sets the rest of the address bus lines to ensure that the system address decoding block sends a signal along the enable (E) line of the memory block, shown in the figure. This activates the block – without it the block remains tri-state or 'unplugged' no matter what happens.

The pattern 0010 is fed to the 4- to 16-line address decoder shown in Figure 2.3. This is an electronic circuit which simply converts each of the possible sixteen patterns on its input lines into a signal along just one of its sixteen output lines. Each output connects to, and activates, just one of the memory locations.

The pattern 0010 thus activates memory location number 2. At the same time, the MPU sends a signal along the read/write line (R/W) to say whether data is to be written into the memory location, or read from it. This ensures that the data direction gate shown passes data in the correct direction. By this means, the MPU can place an address on the address bus and collect the data from the required location along its data bus. Similarly, it can place data on the data bus to be fed to that location if R/W is controlled correctly.

The addressing is thus used to ensure that for the required instant in time, the data bus is connected uniquely to the required memory location, in the correct memory block.

In practice, memory blocks are often single chips with all of the circuitry of Figure 2.3 contained on the chip. In addition, typical current memory densities would be expected to include anything from 2K bytes (eight bits each) to 32K bytes. This requires a large amount of address decoding for individual memory locations, and is an important advantage of having the whole memory block confined to within a single chip, or at least as few chips as possible.

### Memory Types

The two types mentioned above are RAM and ROM. In addition it has been pointed out that ROM is a generic name for several different types of read only memory, of which the EPROM is an example. In practice, the subject of memory is considerably more complex than this, particularly given the latest advances in this field. In a small controller, as we have seen, it is normal to expect some memory space to be devoted to RAM, and some to ROM. In addition, it is usual to use RAM and ROM chips which are as nearly pin-compatible as possible, and thus provide sockets on board for either type. In general, it is necessary to have at least one chip of each type.

There are two main types of RAM in common use – static and dynamic. Static RAM is easier to use and can be provided in a chip package which is pin-compatible with ROM chips. Dynamic RAM chips are more difficult for a designer to use, and they are not replaceable with ROM packages. Dynamic RAM is also much higher in density and cheaper than static RAM, as well as being more power hungry. Further explanations of the electronic differences between these types of RAM are beyond the scope of this book, but in general, dynamic RAM is used where large memory capacities are required, and static RAM is used where comparatively low capacity is needed, but the memory has to be flexible.

The ARC40 board, for instance, has three sockets for static memory chips. These apply to a system of pin-compatible RAM and ROM devices. The sockets here can contain one RAM of 2K bytes and one EPROM of 4K bytes, leaving one socket spare. Another configuration might use two

8K RAM chips and a single 16K EPROM chip. This would be more expensive, but the flexibility of memory configuration means that a lower cost can be achieved when lower memory capability is required. In addition, some control applications may need large RAM and small ROM, or the other way around. This system of pin-compatible static memory allows the memory to be personalised easily and at low cost using just three sockets. It is worth looking for this system in any small controllers which are considered.

The EPROM types which are referred to here are the '2732', '27 128' and '27 256' types. You will see them referred to in these terms in controller literature in general. The numbering arises from the amount of memory which they contain. The '27' part is a generic number which means that the chip is an EPROM of a given pin configuration and a given programming and erasure method. The last part of the number defines the number of K bits which are stored within the chip. Thus, a 2732 has 32K bits of storage, or 4K bytes. It is, of course, the number of bytes which is important to the user of the chip, and you will have to divide the number following the '27' by 8 to find it out. Thus, the 27 128 has 16K bytes of EPROM storage on chip.

In addition to the above types of semiconductor memory is the 'EEPROM' type which stands for electrically (or electronically) erasable programmable ROM. This type of ROM can be programmed, erased and reprogrammed electrically. The difference is that the previously mentioned EPROM is programmed electrically but has to be erased by shining UV radiation into a quartz covered window formed in the top surface of the chip package. The EEPROM is a major breakthrough in the area of ROM, but its packing density, speed and price have still not reached the levels of the 27 series of EPROMs.

EEPROMs (and EPROMs) work by storing data as minute electrostatic charges within the silicon chip. These charges are insulated from their immediate environment, and as such do not leak away quickly. They are expected to last a number of years. This method of non-volatile memory storage is in contrast to classical methods which employ magnetism in one form or another.

A further type of ROM has been mentioned in Chapter 1 – masked ROM. When a chip is manufactured, a form of photography is used to produce a set of masks through which, by various processes, the mask's pattern is impressed on the surface of the silicon chip. Thus, the word 'masked' simply means associated with the actual manufacture of the chip. Masked ROM, therefore, is ROM which is literally manufactured along with the electronics residing on the chip. That is, the binary digits (1s and 0s) of the program or data to be stored on the masked ROM section are irrevocably formed on the chip and cannot subsequently be changed.

Another important branch of semiconductor memory chips is that of the low power types of static RAM. Most RAM chips consume enough power to make them unsuitable for ultra low power battery applications such as are found on the small calculators. A special technology, generally called 'CMOS' is used to fashion low power memories, and such chips are available in a number of forms, including pin-compatible packages to the 27 series of memories.

Other types of memory include disks and bubble memory, which rely on a technique of using magnetism. Such memory devices have the advantages of high density and non-volatility (no loss on switch-off), but are slower than semiconductor memory, and are thus unsuitable for the internal program memory of an MPU system. Such memory is thus called 'peripheral memory'.

*Table 2.1* Memory types

| Memory | Type | Uses and Comments |
|---|---|---|
| Register | Internal to MPU | Used as instruction store, and for general state of the MPU at any time |
| RAM | Volatile random access semi-conductor memory | Internal read/write memory of the system |
| ROM | Read only semiconductor memory of any kind | For permanent non-volatile storage of programs and data |
| EPROM | Erasable programmable ROM | ROM which may be programmed by the user but can only be erased using UV light |
| EEPROM | Electronically erasable pro-grammable ROM | Type of ROM which can both be programmed and erased electro-nically – almost non-volatile RAM |
| Masked ROM | Pure ROM | The binary data stored on this type of memory is manufactured into the chip |
| CMOS RAM | Low power semiconductor RAM | For battery and other low power applications such as battery back-ed up RAM |
| CMOS ROM | Low power semiconductor ROM | For battery and other low power applications |
| Disk | Magnetic peripheral memory | Used for mass storage of programs and data: much slower than semiconductor memory but non-volatile |
| Bubble | Magnetic peripheral memory | Used for mass storage of programs and data: much slower than semiconductor memory but non-volatile |

The term 'core memory' is sometimes used for the internal program memory of a computer system, and is a throw-back to the original types of internal memory based upon small magnetic cores threaded onto thin wires.

To summarise the above, Table 2.1 gives the main memory types referred to, along with their uses and other comments.

This gives some insight into the inner workings of the memory system and the way in which the electronics cycles continuously while running a program. Such considerations are essential to the hardware designer. However, it is worth emphasising that most of the details of the fetch and execute cycle are automatic and not relevant to the programmer – he just expects the system to do all this in the background. In a control system he is more interested in what is going on in the outside world via the I/O lines than the way in which his instructions are being executed.

## Input/output

The programmer is particularly interested in the format of the I/O structure of the system, as it is through this section that he gains access to the workings of his program.

I/O devices are normally connected into an MPU system in a similar manner to that of memory devices. Indeed, most common forms of I/O chip will look just like a set of memory locations to the MPU, and can be treated identically. There are many ways to include I/O lines in a system, and we will look at one of the most common methods, which uses an LSI I/O chip.

The term LSI (large scale integration) simply refers to the fact that modern chips have a lot of devices formed onto them. We thus talk of LSI chips. There are three other somewhat ill-defined terms in common usage, which compare numbers of semiconducting elements packed onto the chip's surface. The complete list is SSI, MSI, LSI and VLSI – small scale, medium scale, large scale and very large scale integration respectively. This forms a crude but workable method of assessing the complexity and number of facilities which a chip offers. In fact, the I/O section of a controller can be formed from any one of these levels of technology. The higher up the scale one goes, the more 'intelligent' and thus easily usable become the I/O chips.

We will start by looking at LSI I/O devices as they are easy to deal with and introduce input/output quite adequately. There is a very large range of LSI I/O devices on the market for applications from simple parallel I/O lines to complex communications devices which include the automatic handling of the most complex of protocols. We will look at a fictitious parallel I/O device which embodies the main features of such chips. At the same time, the basic principles will carry over to any other types of LSI I/O device in general. This will aim to assist the reader in reading any

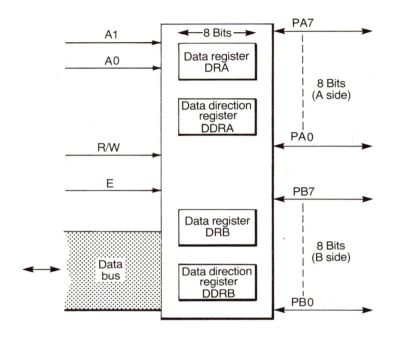

*Fig. 2.4* Parallel LSI I/O chip.

literature which may be met in this area.

Figure 2.4 gives a block diagram of an I/O chip having 16 I/O lines. The chip is shown divided into two separate sides – labelled A and B. Each side controls eight parallel bits of I/O which are labelled PA0 to PA7 and PB0 to PB7 respectively. In addition, each side has a data register (DRA and DRB) and a data direction register (DDRA and DDRB), which are internal memory locations that can be read from or written to via the data bus shown, when the E pin has been activated. The R/W line, as for memory, is used by the internal logic of the chip to determine which direction the data is to travel on the data bus.

A general parallel I/O device will allow the programmer to define the direction (in or out) for each I/O line on the chip. This information has to be written to the chip, and the chip is then expected to adjust the I/O lines accordingly. The direction of each line, therefore, has to be stored somewhere, and hence the DDRs. Each of the eight bits of the DDRs defines the direction of one of the I/O lines on its side (A or B). Thus, for instance, an 8-bit word might have been sent to DDRA as follows:

| DDRA | 1 | 0 | 1 | 0 | 0 | 1 | 1 | 0 |
|------|---|---|---|---|---|---|---|---|

We will assume that the logic level '0' signifies out, and logic level '1' in, and that the extreme right hand side of the register contains the least significant digit. Thus, the above number defines the following I/O pattern:

| | |
|---|---|
| PA0 | Out |
| PA1 | In |
| PA2 | In |
| PA3 | Out |
| PA4 | Out |
| PA5 | In |
| PA6 | Out |
| PA7 | In |

Each output line may be connected to an amplifier of some kind, and will be able to switch an external device on or off. Each input line will be able to read the state of an external switch. We will see examples of the types of device which can be connected to these lines later in the book. For now, it is clear that the outside world can be sensed and controlled to some degree by this device.

The DRs are used to send data to and from the I/O lines. As for the DDRs, each bit in each register applies to a corresponding I/O line, and the DRs provide the MPU with a direct method of reading and writing to the I/O lines. For instance, if all the lines of the A side are set to be outputs, then sending the following binary number to DRA:

$$1 \quad 1 \quad 1 \quad 1 \quad 1 \quad 1 \quad 1 \quad 1$$

will switch all the A side output lines to logic level '1'. Similarly, if the lines are set to input, and all the I/O lines were receiving a '1' level, then DRA would contain the above binary number and the MPU could read it. Thus, all I/O transactions occur through the bits of the data registers.

In order to allow the MPU data bus to be connected to the correct internal register, the lines A0 and A1 shown are used. These are actually connected to the two lowest address lines of the address bus. The four different patterns which are possible on these two lines will select, via internal logic, the four different internal registers. Thus the usual address decoding block of the system signals to the E pin of the chip, and then the lowest two address bus lines select the specific internal register to be contacted at any time.

This gives some idea of the way in which electronics and software meet. The software is concerned here with setting the I/O chip up at the start, and then using it to read and write to the I/O lines via the chip's internal registers. These tasks are common to most LSI I/O devices, even those which do not primarily have parallel lines to control.

In order to illustrate that LSI devices are not essential for I/O, we will have to become slightly more technical than has been necessary so far. Figure 2.5 shows a pair of chips which give eight inputs and eight outputs.

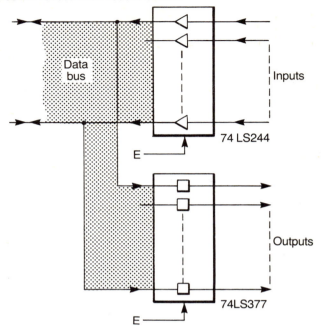

*Fig. 2.5* SSI chips in I/O.

This uses a couple of SSI chips which are fairly straightforward to describe. The chip type numbers shown are standard to the specific chips described, and may be manufactured by many different companies, though different manufacturers may add their own suffixes or prefixes.

Figure 2.5 is a typical chip-based circuit. The chips themselves are indicated by rectangular boxes, within which are shown the elements which connect to the external pins. In a real circuit, pin numbers would also be included on the lines shown going in and out of the chips.

As before, each chip shown has an E pin. The 74LS244 chip contains eight tri-state 'buffers'. Their function is to pass logic levels from their inputs to their outputs when enabled, and otherwise to remain in the tri-state (or 'unplugged') mode. Thus, even though their outputs are connected directly into the data bus, they cannot interfere until enabled, and at that time the MPU is specifically waiting for a reply from that chip. Once again, it is important to remember that anything which is capable of outputing to the data bus must be capable of being tri-state until enabled. It is the electronic logic of the address decoding block which will signal to the E pin of the chip.

In the same way, the E pin of the 74LS377 is also connected to the address decoding, but the function of this chip is to 'latch' the logic states of its inputs onto its outputs. Thus, when the E pin is pulsed, whatever binary pattern is present on the chip's inputs appears on its outputs and stays there until they are set to a new pattern next time. This chip does not need to be tri-state as it does not feed onto the data bus (see Figure 2.5) and so cannot interfere. It is important for output lines to latch, or remember, their logic states, because the MPU only applies the required output pattern for a fraction of a second, and then it continues with other tasks, and the data bus changes accordingly. The correct output pattern has to be accepted and placed on the output lines 'on the fly' as it were. It is up to the MPU, of course, to ensure that the chip's E pin is enabled at exactly the correct time to catch the required pattern from the data bus.

This shows the difference between an LSI I/O system and a 'discrete' I/O system which just contains ordinary electronic logic chips. The LSI device is a single chip which allows sixteen I/O lines, each of which may be set separately to in or out as required. The MPU communicates with it as if it were a set of four memory locations. The discrete device, on the other hand, requires two separate packages, with all the attendant interconnections, and it only supplies a fixed eight in and eight out. No flexibility on these numbers is possible.

At one time, it was lower in cost to design using discrete devices if the application allowed it. However, with the extreme decrease in LSI chip costs, it is no longer true, and the added inconvenience of having two chips to connect often makes the LSI device more attractive.

### Peripheral functions

This is the last block to explain in Figure 2.1. It lumps together all the optional additional devices which one may find in a controller. For instance, as mentioned above, it may be necessary for a given application to store data which does not disappear upon switch off. This could be

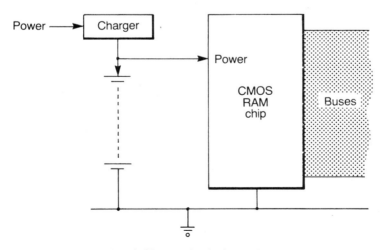

*Fig. 2.6* Battery backed up RAM.

achieved using a block of EEPROM, or by using a battery backed up RAM section. Figure 2.6 illustrates a block diagram of the latter case.

A battery charger, permanently connected to the system's power supply, keeps a rechargeable battery on charge. The battery in turn feeds the power supply of a CMOS RAM chip (or chips). Any information which the programmer wishes to remain intact after the power supply is switched off, or if it fails, is stored in this section of RAM. When the power is off, the battery keeps the contents of RAM intact. A CMOS RAM will allow the battery to keep the memory powered for a considerable time. As a comparison, a small battery in a calculator, which uses CMOS RAM, may last for months or even years of intermittent use. In addition, the solar powered calculator is quite common and shows how little power is required for some CMOS circuitry.

Battery backed up RAM is an alternative to EEPROM where the normal speed of read/write memory is essential. However, it does require a battery, and is a somewhat cumbersome addition to any controller. If a few variables are required to be stored very rarely, then EEPROM is the answer. The type of example may be found in the calibration of a laboratory instrument. In this case, the calibration is rare and the operator does not mind a short wait if there is a lot of data to be stored. Having stored the data in an EEPROM, this data does not need a battery to keep its memory stored, and in many ways it is an ideal solution to the problems of producing non-volatile RAM.

Another application of non-volatile RAM is for a data logging instrument which must be energised for a short time, take some readings, and then power off. This might be in a rugged environment where the batteries cannot be changed very easily. This intermittent usage saves power. EEPROM is ideal for storing simple counts and remembering them until the unit is powered up again to continue counting from where it left off.

The main disadvantage of EEPROM remains its programming needs. If this is not a problem, then EEPROM is an excellent alternative to battery backed up RAM or magnetic memory.


## THE OVERALL PICTURE

As can be seen from the above, a controller is a fairly standardised device depending upon a set of basic modules. It would be difficult to discard any of the modules given. However, it may be that many or all of the modules can be confined to a single chip, or they may range over a number of boards. As we have seen, the exact construction and design method is dependent upon the facilities and attributes required of the finished system.

It is most important, when considering a project, to decide on the level of technology which will be most applicable, and this depends upon a number of different parameters. For instance, a high volume cheap controller will use a single chip micro; a low volume, low cost control application where there is little money for research and development will use a single board controller with integral development system, and so on. An understanding, however superficial, of the technology itself is of considerable help in ensuring that the project is approached correctly – even if an outside consultant is used from the start. It has been proven time and time again that it is far more efficient if such a consultant can talk to a knowledgeable recipient of his services, rather than someone who knows nothing of the technology.

Three main methods of construction are used for the majority of microelectronic controllers. These are single chip, single board and multi-board controllers. We have looked at the first two, and the next section looks at the third alternative.

## MULTI-BOARD CONTROLLERS

In some instances, the control problem is so large and complex that it may be an advantage to use a multi-board micro system. In some ways this construction method is becoming less and less attractive as the complexity and sophistication of integrated circuits increases. However, it is important to realise that this is an alternative, and it has a number of advantages.

The philosophy behind this design is to confine like modules to a single board and fit all the boards into a system, perhaps by a system of buses. For instance, the memory requirement may be sufficiently great to require a complete board all to itself. Similarly, I/O may require a complete board of devices for its implementation.

A major advantage of the multi-board approach is that the controller can be assembled from a set of simple parts to produce as complex or simple a system as is required at any time. This theoretically allows infinite upwards expansion of I/O, for instance.

There are many standard systems of interconnection in existence, and they are simply bus systems just as described in the above sections. The typical form of a multi-board system would be a set of standard printed circuit boards which plug, via edge connectors, into sockets in a back plane. Figure 2.7 shows a sketch of the situation. Here, just three sockets are shown on the backplane, with two boards in place. A power supply unit (PSU) is connected to the backplane, and hence to all the boards, via a set of PCB tracks.

The backplane is simply a printed circuit board which electrically connects all equally numbered pins on all the sockets to each other. A typical range of boards might include a central processor board, which may be capable of acting as a single board controller for simple applications. To expand the system, extra memory and I/O boards would

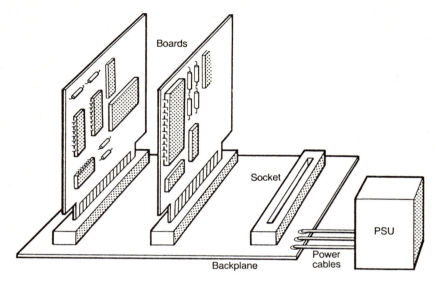

*Fig. 2.7* Multiboard system.

be assembled together to fit the situation. Some specialised boards may also be offered in a typical proprietary range, by a given manufacturer. There may be analogue boards dedicated to allowing continuously varying voltage levels to be input and output from the system. Another board may offer dozens of separately programmable digital I/O lines to allow a large control function to be implemented.

Such multi-board systems may be found, for instance, in a large data collection system where many machines on a shop floor are to be continuously monitored and reported upon.

The definition of which line goes where on the backplane is one aspect of the use of these systems. It is usual for any board to be able to fit into any socket, which means that the line numbering definition is important. Another aspect of the definition which is also crucial is the exact definition of the electronic states of each line at any time. This is important for a bus system which is to be open to any manufacturer to add to. There are systems which are unique to a given manufacturer, but most useful bus systems will allow any one of several manufacturers' goods to fit into the system. This definition of the bus system was often the downfall of early buses where the definitions were less than complete.

Some typical systems are:

| | |
|---|---|
| The S100 system | – which is somewhat obsolete |
| Multi- bus | – which turned out to be rather too complex for most applications |
| STD-bus | – which was common in the USA, and not so common in the UK |

| VME-bus | – has become fairly popular |
| STE-bus | – which is fast becoming a standard and is simple enough to be used with great ease |

In addition, many companies also have their own bus systems, and in many ways the exact position with respect to the bus system employed in a given control situation is a little academic. The important point is that there are plenty of boards available for the system. It is also worth considering whether one is confined to a single source, which may decide to stop making this particular range in the future or even go bust leaving the users with an obsolete system. These points should be borne in mind carefully whenever a multi-board system is proposed.

As explained above, it is often possible for a small SBC to be expanded by including it in a larger bus-orientated system. This is true, for instance, of the Arcom ARC41 series of boards, and this range includes a number of specialised I/O and other boards as described above. This may also be something to look for when an SBC is proposed if there is any chance that expansion may be necessary in the future.


## A LARGE CONTROL SYSTEM

As promised earlier, we will now look at an example of the overall analysis and design of a system. Since we have spent some considerable time so far on small controllers, and mostly within the product control side of industrial applications, the example below concerns a larger project, one nearer to a typical process control system.

Process control systems can take many forms. There are simple compact systems which perhaps just monitor a light cell on a conveyor belt and activate an air blower to remove an apparently misformed product when it comes by. Further up the scale, there is the type of system which might control the dispensing and mixing of a number of chemicals in a vat, and then precisely dispense the finished product into cans of a given size. At the top of the scale there is the large computer system approach to a whole shop floor or production hall, where all the processes are controlled and reported upon in real time, and buttons can be pressed to find the exact state of the shop floor at any instant. This last example may be termed production control, rather than process control, but the overlap is considerable.

We will look at a somewhat different application from any which has been mentioned before. It is an important one in manufacturing, though it has yet to be fully utilised by the industry. It is the application of 'data collection'. Though this is not strictly a control application, in that it is mainly concerned with observation, it can form the first stage in a much wider and more encompassing form of control if taken to its ultimate conclusion. Once the process is fully observed, it is then possible for the computer system to make decisons and issue instructions to ensure that the system remains efficient, and the production plan is being implemented. In this manner, the computer effectively runs the shop floor, and thus controls the overall process. In addition, the various parts of a data collection system are controllers in their own right, and as such they form a typical example of the technology discussed above.

## A data collection system

A manufacturer wishes to observe the workings of his shop floor in order to see how efficiently his process machinery is being used, and whether his production plans are being adhered to. At the same time he wishes to take in data of the timings of various jobs in which his personnel are engaged, and produce readouts of the bonuses he has to pay them automatically.

The exact nature of the machinery required here depends upon a fairly complete specification of the problem in hand. In fact, this is the first stage of any project and should be made as accurate as possible before any work is started.

We will look at a fictitious specification to make the main points in this section. Suppose the shop floor consists of ten machines. These machines are five plastic injection moulding machines, three cutting presses, and two finishing tables where groups of operators assemble, deburr plastic mouldings, pack and perform other hand operated processes. The object is for a control system to monitor the machinery and the production process, report on progress at any time, and calculate bonuses.

The mix of operations in this shop floor into largely automatic processes such as injection moulding, semi-automatic processes of the cutting presses and fully manual processes is not atypical. It is not easy to blend all these activities together into a micro system, and predictably the greater the manual part, the more difficult the problem.

The first task in the systems analysis is always to determine the inputs to the system and its outputs. The manufacturer probably has a good idea of what he wants, but this will be held in his head in a non-computer-orientated manner. The systems analyst has to turn this into logical sequential operations, identify the overall inputs and outputs, and decide on how to receive data from the shop floor and how to present it to the production managers.

## The specification

In this case, the manufacturer declares that although the first system to be put onto a computer system is for the small shop floor described above, he has a number of new machines coming, and several other shop floors in his works, all of which he wishes to tie into the system at some time or other. Some of these machines are to be monitored for continuously variable

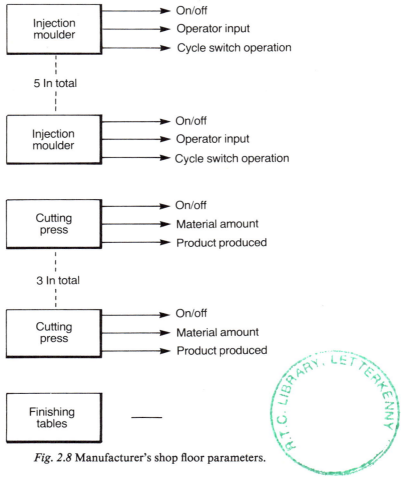

*Fig. 2.8* Manufacturer's shop floor parameters.

parameters such as length, weight, and so on.

However, in the first instance, it is decided that the parameters to be taken in by the system will be as shown in Figure 2.8. The injection moulders will be monitored for two parameters automatically. Firstly, each time the machine is switched on and off, say at the start and end of a session, this must be noted by the electronics. Secondly, a signal must be given each time the machine cycles. These are taken directly from the internal power switching within the machine, via some interfacing. Thus the computer system will always know when the machine was turned on or off, and it will have a constant record of the number of cycles performed in each 'on' period.

At the same time, the operator will be required to input a unique operator number into the machine, along with a job number, each time he or she operates the on/off switch – otherwise a loud buzzer sounds as a reminder. Similarly, the system would be monitoring the actual operator and job at any machine, and if it decides that the operator is trying to operate the wrong machine, a similar buzzer sounds. This gives a small degree of control over the operator's actions. If a buzzer sounds, it is up to the operator to put the problem right, or a supervisor will be required to sort out the problem.

The cutting presses are controlled in a similar manner, except that the operator will be required to input the amount of material he uses during a given session and the amount of product produced. This is not too satisfactory, and it is up to the operator to ensure that he draws a given amount of material from stores and checks on the amount used before finishing his session on the machine. This is typical of the less automatic machinery. The efficiency of the cutting press operation depends upon how the material is laid in the press, what sort of tool is used, and so on. As before, the operator inputs his number as well as that of the job. However, with the injection moulders, the amount of material used and the amount of product produced are dependent upon the number of cycles of the machine.

The finishing table is difficult, expensive and extremely fiddly to monitor or control by computer. Most maufacturers would accept this and realise that, if nothing else, this is the last place to attack – it is better to leave it until absolutely necessary, if ever.

This indicates the inputs which the system will take in, and if you have some experience of this type of manufacturing, you will be able to see a number of different deductions which could be made from this set of shop floor statistics. For instance, it would be straightforward to keep a background log of the total number of working hours each machine clocked up, and automatically signal the need for given types of maintenance. This is called automatic planned maintenance. Another deduction which could be made would concern material wastage. This would also need, theoretically, the input of material quantities issued to the shop floor, as well as total product leaving the floor. However, even without these it is possible to see trends of material usage from operator to operator, and thus the efficiency of each operator.

Figures can also be assembled for percentage machine usage time – this can throw up information as to how efficiently a given machine is being used, and whether it is overworked and in need of further machinery of the same type. However, this is a little intangible and will be a consequence of the production controller's conception of what he wishes the shop floor to be producing under each job at any time.

The main output required from the system, as indicated above, is the current state of the jobs on the shop floor – i.e. specific progress of each job. In addition, machine down time as a percentage of the day is required, as well as the exact hours and number of units made for each machine operator to calculate his or her bonus.

This completes the overall look at inputs and outputs, and in a real example these parameters would be agreed exactly and described in the specification document in detail. Such details are often best put together by a computer expert as they must be written in terms which will transfer to the computer. Also, a good analyst will be able to recommend further outputs which may be produced for little extra programming effort, given the data which is already being taken in.

In this section, the analyst has identified the current I/O requirements specifically, and also ensured that he has been made aware of the future expansions of the system. There is nothing worse than designing a system from scratch only to find that the user requires it to expand to five times its original size within a year. A good analyst will find this out at the start, although he will also point out that allowing for great expansion at the start is an expensive business, and the user should be quite sure of what he is doing.

We will now look at a typical set of operations which the systems analyst/designer must go through to specify the system correctly. This should give the reader some insight into the main decisions to be taken while considering the use of a control system of any type.

## I/O Considerations

We will look, in a later chapter, at I/O devices specifically, but we have

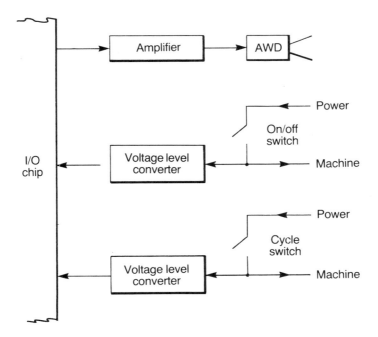

*Fig. 2.9* Machine I/O.

seen enough so far to discuss the inputs from the machines and how this may be achieved efficiently. The operator interface is also the subject of a later chapter and we will not look at this in fine detail here.

The monitoring of the on/off switch of the machine is very simple. It consists of a connection wire to the machine side of the switch, a voltage level converter to the levels of the control electronics, and an input line of the type shown in Figures 2.4 and 2.5. The cycling of the injection moulding machines is detected by a similar method. A switch within the machine is monitored to detect its switching action, and this is also transformed and fed to an input line.

The buzzer line is a simple output from the controller connected to an amplifier and an audible warning device (AWD) of some kind.

These I/O lines are shown in Figure 2.9. The voltage level converters may simply consist of rectifier diodes and some passive components to smooth the DC voltage and divide it down to logic voltage levels (+5 V and 0 V). This uses three I/O lines per injection moulder. The cutting presses would only have two as the cycle switch is redundant. This gives a total number of I/O lines as fifteen for the injection moulders and six for the cutting presses, giving twenty-one in all. However, this is only a small

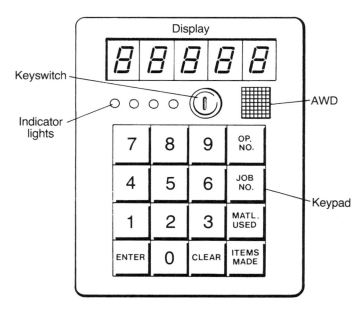

*Fig. 2.10* Operator interface front panel.

part of the I/O – there is still the operator interface to consider for each machine.

Figure 2.10 shows a possible operator interface front panel for each production machine. It is a rudimentary keypad and display system, and allows the input and readout of numbers primarily. In addition, some indicator lights and a keyswitch may be included to make the system more flexible.

The display has several seven-segment calculator type display digits. Some simple letters can be read out here, such as lower case 'b', and upper case 'E'. There are enough letters possible to allow very simple and stylised error and other messages to be put up on the display. However, the display is primarily to allow the operator to check his input.

When the operator inputs a series of numbers via the keypad, they appear on the display as with a calculator. If he is unhappy with the number shown, the 'CLEAR' key will clear the display and he can try again. When satisfied, he presses the 'ENTER' key to indicate this to the computer system. The operator only inputs one of four things, and these are indicated by dedicated keys on the keypad, in order to simplify the operator's job as far as possible – which is essential in any system of this type. For instance, when he wishes to input his operator number, he presses 'OP. NO.' and the computer activates the keypad. He inputs the number, presses ENTER, and the numeric part of the keypad is then inactivated until he presses one of the four command keys again. If he keys in an impossible or unexpected number, the AWD sounds, and he is allowed to correct his input, or the supervisor arrives to find out what is happening.

This type of operator interface requires an estimated sixteen lines of I/O to operate it. This means that a further 128 I/O lines are required by the eight machines in the system. The total number of lines is now around 149. This cannot be catered for by a simple SBC, unless some expansion of the facilities is employed.

There is little else needed to and from the shop floor, and the next step is to design the complete data collection system.

## Overall design

There are two ways to attack this type of system. Firstly, a central computer system could 'talk' along standard communications lines to nodes at the machines, each of which would be a microcomputer in its own right. These computer-based nodes would have a front panel as in Figure 2.10, as well as interfaces as indicated in Figure 2.9, and run the entire system with ease. The nodes would be intelligent enough to collect data, make decisions as to the integrity of data input, collate data and send it down to the central system when requested to do so. Such a multi-computer system can be made to perform almost any type of data collection and process control task.

Secondly, a central control system could be produced with enough I/O to connect to each of the operator interfaces directly, and take on the task of monitoring the entire system by itself. This is the single-computer approach.

The first case above is complex and expensive as each node has to be programmed and tied together in a network. The second case has the advantage of being easier to develop, but is slow because a single processor has to do all the work in the system all the time. However, this is not difficult for a controller, and there is no reason why a single processor should not be sufficient. The only problem to overcome is to ensure that the controller is fast enough not to miss a single cycle of an injection moulder, or a single keypress by an operator – even if all such events occur simultaneously.

The speed is not, in fact, particularly important. An operator does not mind waiting a short time for the system to respond, and the cycles' inputs can be latched (or remembered) into an input circuit when they occur, and thus not missed if the processor is engaged elsewhere. The only proviso is that each latched machine cycle signal must be read and cleared before the next one arrives from that same machine, or that next one will be lost. Again, given the natural speed of an injection moulder, this is no particular problem. In fact, it is the normal experience of this type of system that even when production is working flat out, the processor will only be utilised for a small percentage of its time. This leaves plenty of time spare for such tasks as keeping a count of minutes, hours and days to provide an internal real-time clock function.

If the single-computer approach is adopted, it would be appropriate to

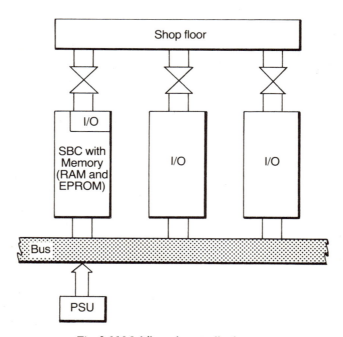

*Fig. 2.11* Multiboard controller layout.

employ a multi-board system for the computer. Again, this could be based around a standard expandable SBC. Figure 2.11 suggests a board layout. Here, the main computer system is confined to an SBC, with some I/O already available. The other boards simply supply more I/O lines. These lines would probably have to be interfaced to transmitters and receivers to allow the logic voltages to be transformed to and from more suitable levels to be connected throughout an electrically noisy environment, perhaps for several hundred feet. This is a crucial consideration, and much of the electronics involved in any project of this sort is to ensure that electrical noise does not interfere with the system. However, this can always be overcome, and should not prevent data from flowing correctly around the system.

There are three main advantages to choosing a multi-board system, as opposed to simply expanding the I/O of an SBC, perhaps with a specially designed expansion board. Firstly, the research and development is considerably easier and cheaper. Secondly, the manufacturer's stated future expansion plans can be catered for simply by expanding further, using standard modules. Thirdly, if an I/O line goes down at the computer end, it is simple to remove boards one by one and replace with spares until the fault is remedied. The offending board can then be sent back to a specialist for repair. In other words, the more standard and modular the system is, the easier it is to maintain.

There is, of course, a limit to the expansion possible with an SBC and a bus system, and it may be that the full aspirations of this particular user will require several computer systems. However, with a multi-board system it is an easy matter to customise the hardware to the requirements exactly. It is also not difficult to arrange communications between computers to share information. This is largely a matter of function – i.e. software.

In fact, there are several points which a purely hardware description cannot cover. We will now look at the general nature of the software tasks.

## Software

The software for this project would occupy what is known as a 'polling' loop. This means that the software would be executing a set of tasks circularly, and interrogating (or polling) the input lines sequentially, looking for input. For instance, four of the keys on the keypads are always active. It is up to the software to view these keys, continuously. If one of them is pressed, the computer then watches for numbers being input. As they are input, the computer reads them, and echoes them back to the display. The computer meanwhile watches for a 'CLEAR' or 'ENTER' and acts accordingly. However, at the same time, the software must not forget to watch continuously for more operator inputs and cycling of the machines.

There are two common ways of achieving this state. The first is simply to execute a sequential loop which allows a certain amount of time to each input line in the system. This is quite possible and, given the comparatively low speed of the individual items, the system will work satisfactorily. However, it is somewhat difficult to program. It would be very useful if the hardware could be utilised to give some tasks a higher priority than others. For instance, suppose the machine cycle and on/off switch inputs could interrupt the processor no matter what it was doing. These could be recorded, which is a very quick and simple task, and then the processor could simply return to its interrupted task as if nothing had happened. If this were possible, the programmer would be able to split his tasks naturally into two separate areas. These would be the operator inputs and the machine switch inputs.

In fact, all MPUs allow this type of interrupt, and to see how it works we will now look at this interesting and useful attribute.

## Interrupts

As explained above, the MPU's registers give the complete current state of the processor at any instant. In particular, the 'program counter' register contains the address of the next instruction to be processed. This is the way in which the processor keeps track of its position within the program. If you could change the program counter contents, the processor would blindly follow the new contents, and fetch the next instruction from that address. This is arranged by a special technique of interrupting.

All MPUs have one or more special interrupt pins. When one of these pins is brought to a low level (logic 0) typically, a special sequence of events occurs within the MPU automatically. These actions are indicated by Figure 2.12.

The first action is to save the current program counter onto a special memory region called the 'stack'. The program counter is then filled with a new address called the 'interrupt vector'. The value of this address can be decided by the programmer. At this address, it is up to the programmer, or hardware designer, to arrange that there is a special routine called the 'interrupt routine'. This is then executed, and when finished a special 'return' instruction causes the MPU to retrieve the program counter from the stack, where it was left. The MPU continues with the original program as if nothing had happened. This process of interrupting can occur anywhere in the main program, and has no effect other than to slow it down. The interrupt routine, which is executed every time an interrupt is demanded, is a truly background activity of which the main program is not 'aware'.

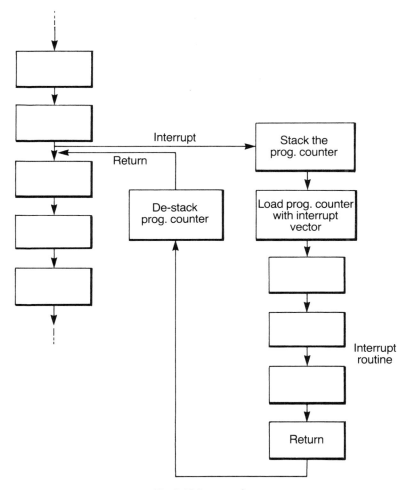

*Fig.2.12* Interrupting.

## An example of the use of interrupts

A standard application of interrupts is in the production of a real time clock from an external oscillator. If a regularly cycling oscillator is connected to the interrupt pin, then interrupts occur at a known time interval and these can be counted and used to update seconds, minutes, hours, etc. in a real time clock.

A typical setup would be to arrange an oscillator to interrupt the main program every tenth of a second. The interrupt routine would be counting tenths of a second, and storing the counts in a memory location. Every ten of these, it would increase its count of seconds held in some other memory location, and every 60 of those its count of minutes, and so on. A group of memory locations would thus contain the accurate real time to within a tenth of a second at any instant.

The main program may have a use for real time, as does our data collection system above, and it could simply read the updated time at any instant, which is kept up to date by this hidden interrupt process.

You may imagine that executing this interrupt process every tenth of a second will have a significant slowing effect on the main program, so we will now estimate that effect.

If the MPU is cycling at around 4 megaHertz (megacycles per second) then each cycle takes a quarter of a microsecond, or 250 nanoseconds to complete. If an average instruction takes around six cycles to complete, then it takes 1.5 microseconds. The interrupt routine is not complex and it might involve a few dozen instructions in a simple counting routine, but we will be conservative and assume 250 instructions for this task. Thus the total time for running through the interrupt routine is around 375 microseconds. There will be a few more microseconds at the start and finish to stack and de-stack the program counter and other registers if appropriate, and the total may be, with a margin, 400 microseconds in all. Some MPUs will take more time, others considerably less, but this is a general estimate.

Thus, every tenth of a second, the MPU wastes 400 microseconds. A total of 4000 microseconds is wasted per second (which contains one million microseconds). This gives a percentage waste of just 0.4% of the MPU's total time which is quite irrelevant in most systems where the effective utilisation rate of the MPU may only be a fraction of a percent anyway.

**Interrupts in the data collection example**

In the data collection example described above, it is not proposed that a real time clock should be produced by interrupts – it is normal to expect the processor board in the system to contain circuitry for a real time clock already, with its own internal counting chip. Such a chip would be connected into the system as a group of memory locations, and reading its internal registers will give the real time at any instant.

The application of interrupts suggested here is to help make the system easier to program, and give some sort of priority to the cycling and switching of the machines. They will not wait in the same way as an operator can for the computer to service his input.

The only part of the machine's switching which might need to be recorded quickly is the cycling. It will be assumed that it does not matter if it takes even a few seconds to read whether the machinery has been switched on or off, for instance. Thus there are five injection moulding machines with cycle switches which must be monitored by interrupt. This means that there are five possible interrupting signals which could come in simultaneously, none of which can be ignored or forgotten. To cater for this problem, a special interrupt priority circuit is required, and again most processor boards would have such a circuit included – if not, a special interface can be designed easily to fit between the machine and the processor board.

When a machine completes a cycle, it sends a pulse to the computer. This pulse interrupts the program, and the interrupt routine simply increases its count of the cycles from this machine and returns to the main program. Thus the current count of cycles from each machine is always available to the main program for presentation to the managers.

If more than one cycle pulse is received at the same time, the interrupt priority circuitry will memorise all the incoming pulses, and the processor will increment all the signalled counts before returning to the main program. The interrupt routine simply reads the interface circuitry to determine which machine has just cycled, and thus it ensures that it increases the correct counters.

By this means, therefore, the main program always knows the state of the machine's cycles. Its normal polling operation simply cycles around all the keypads, displays and on/off switches noting changes and servicing operator requests as it can. No machine cycles are lost, and the operator will find that he rarely has to wait a significant amount of time.

In many ways this gives the most efficient use of the technology, and is the most straightforward and compact to develop. Of course, there will be a large amount of cabling to be dealt with to and from the shop floor, but most manufacturing concerns are already adept at this type of enginering.

### The software for the project

The basic controlling software for the project, therefore, is broadly as shown in Figure 2.13. The main loop simply polls around the system acting on keypad pressings and noting times of start and stop of the machinery. There is a fair degree of programming in this main loop; it has to service, apparently simultaneously, eight keypads and displays, for instance, as well as logging the states of the machinery. This could also involve the use of a large amount of memory which can simply be added as an extra board if required. When the loop is interrupted, the interrupt routine updates the cycle counts of the machinery, and clears the memory of each interrupt which has been collected to ensure that no erroneous multiple interrupts are recorded.

The software is purely a control loop. It prepares the memory of the computer for other analysis programs which provide reports on the current state of production, the amount of time each operator has spent on each job and, if the bonus rates are also fed in, it will perform the fairly simple bonus calculation and provide a full printout of this information.

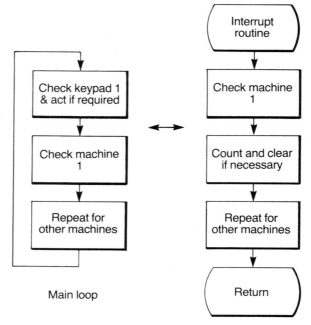

*Fig. 2.13* Overall software of the data collection system.

The most succesful philosophy to adopt for analysing the collected data is to use a second computer as a data processor. This is not a particularly expensive route in this day and age and, for a simple pilot study, it should even be possible for a cheap personal computer to fulfil the need adequately. The data collection controller would feed data along a standard communications line when necessary, and this would be used to produce bonus printouts and general reports on the production data.

Figure 2.14 rounds up the complete data collection system. The machines and operator interfaces are connected to voltage level changing circuitry, which will not be complex. This is then connected to the data collection controller which simply acquires data from the shop floor, as well as controlling it to a small extent initially via the audible warnings and perhaps error messages on the displays. It sends its data and receives its instructions from a computer which is used as the data processor. This data processing computer uses a standard printer for its reports, as well as providing data directly via its screen when commanded to do so.    Of course, none of the above is sacrosanct, and there will be many other system philosophies which could be argued as being applicable in a given situation. However, the system provides a useful benchmark against which other systems can be compared.

Costs are not easy to give in general before a full specification is produced, but the hardware itself will not be expensive in comparison to

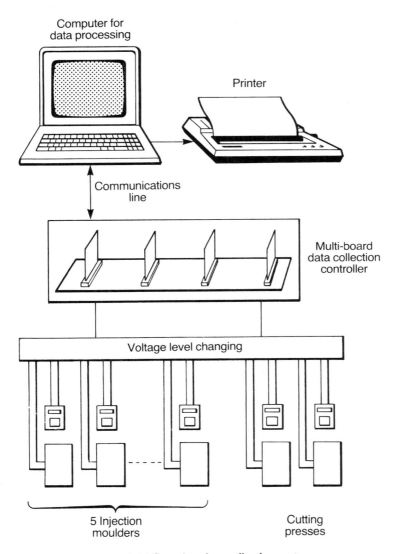

Computer for
data processing

Printer

Communications
line

Multi-board
data collection
controller

Voltage level changing

5 Injection
moulders

Cutting
presses

*Fig. 2.14* Complete data collection system.

other systems. The greatest cost is always the development, and if a given
computer company has used the above system a number of times before,
they will be able to offer more competitive rates than someone attacking it
for the first time. However, the electronic principles involved are simple
throughout, and the software is confined to a single control program in the
controller and some data processing in the computer. It should also be
pointed out that the data processing computer shown in Figure 2.14 is not
necessarily the end of the system. Future expansions can allow for

communication of the data to other systems concerned with payroll, accounts and invoicing, production planning, and so on. In a fully integrated system, all activities would ultimately be recorded automatically from operator clocking-in through to the planning of production, stock control and customer invoicing.

It is also possible to consider using existing computer machinery within the company for the data processing computer part of the system, if there is spare capacity. However, such installations should never be considered for the controller section of the project. The controller must be on-line continuously, and process the information as it occurs – in real time, as it is termed. Such activities are quite incompatible with most larger non-dedicated systems, and should always be left to a small MPU-based system as described.

## CONTROL IN GENERAL

The example above has concentrated on the process control angle, but it is important to realise how similar the main decisions in the design process are in other forms of control. For instance, if a small low cost high volume product is to be automated, it is still essential to tie down a specification of the inputs and outputs first. The number of I/O lines and the electronic nature of these lines must be considered before any technology is chosen. The next step is to consider the function in detail, and determine whether interrupts are useful for the system. This decides most of the hardware which is required for the final design, and it is necessary to ensure that a technology is chosen which will be within the cost/benefit parameters.

For a low cost high volume device, it is not always definite that a single chip device will be the best design, particularly initially. The prototyping and development of a multi-chip or SBC-based design can be more cost effective given the high intitial costs of a single chipper. For instance, it is perfectly possible to put together a cheap and easily developed microcontroller using a single chip MPU which has no internal ROM and thus is available off the shelf. We have already discussed the Z8 in this context. A few cheap chips on a PCB may be the best way to produce a prototype quickly and easily, and then produce an initial production version.

On occasions, of course, this is not possible. For instance, in a case where a high volume hand-held instrument is proposed, there may be no room left after the sensors and other electromechanics have been included in the package. Even here, however, it may be possible to go to other techniques of manufacture, including 'surface mounted devices' (SMD).

SMDs are designed to allow the packaging of large quantities of discrete chips into a remarkably small space. The requirement arises from the fact that conventional DIL packages are many times larger than the size needed for the actual silicon chips within. There is now a strong movement towards producing chips with packages which are only a little larger than the silicon chips themselves. This produces plastic packages which are too small for humans to handle effectively, but automatic machinery exists for their assembly onto PCBs. These packages do not have conventional pins, but rather small metal pads on the outside of the packages which connect to surface pads on the PCB – hence the name SMD. The chip and PCB pads are soldered together by a number of different means, including being coated with a solder paste and baked in an oven to melt the solder.

This construction method produces a development cycle which is many times cheaper than single chip technology, and is highly recommended for medium to high volume work. However, the tens of thousands, and upwards, level of production can still not be beaten by anything other than chips which include the program, and everything else if possible, on the manufactured chip itself. The unit cost of such an approach is very low indeed provided the quantity is large enough.

# Chapter Three
# Sensing and Interfacing to the Outside World

## INTRODUCTION

The previous chapters have dealt with the type of technology which is generally found in the design of the controllers themselves. This chapter concentrates on the technology of the sensors and their interface to the system.

There are many books which describe sensors in great detail, and such data will not be reproduced here. This chapter will describe an overview of some common sensing techniques, and then move on to interfacing techniques such as voltage level changing, analogue voltage sensing, multiplexing to increase the number of I/O lines available to a system, and so on. The aim, once again, is not to give detailed electronics, but rather to explain how common devices work, and how they might be used.

Later in the chapter we will look at some of the more out of the way sensing techniques, in order to ensure that the reader is prepared for situations which may require slightly more devious routes to the collection of information from the outside world. Examples will be given throughout of the use of sensors so that similar situations encountered in the future can be viewed from experience.

## SENSORS IN GENERAL

The human body has a remarkable array of sensors, and the majority of us rarely consider the background mechanisms which give rise to this continuous data gathering. In an industrial situation, a human operator can sense something wrong with a process in an almost supernatural manner, from subtle and even unconscious clues within a process he knows well. The study of industrial sensors is a rather more crude subject in comparison.

Sensor choice normally requires an exact definition of the parameter to be sensed, as a start. For instance, if the temperature of a given object must be kept within certain limits, some form of thermostat would be used. This can take the form of a simple bi-metal strip connected to a switch, and electrical power is broken to a bank of heaters when the given temperature is reached. However, this is only possible under certain circumstances. For instance, suppose the heating is being performed by the burning of solid fuel, and the material being heated is steel in a furnace. The above mechanism would hardly be applicable.

Just as there is a given starting point in the choice of a controller for a control application – namely the number of I/O lines – there is also a starting point in the choice of the correct sensing device. This starting point is simply a complete definition of the parameter to be sensed – including its range, and perhaps how that parameter is derived or controlled.

## PARAMETERS TO BE SENSED

It is difficult to give a list of all the parameters which are to be sensed in any given situation, but a list of the most common would be something like the following:

● Temperature
● Electrical current
● Voltage
● Electrical power
● Pressure
● Fluid flow
● Position
● Movement
● Velocity
● Acceleration
● Proximity
● Strain and force
● Vibration
● Weight
● Level or volume
● Light-based parameters
● Radiation
● Humidity

These are mainly industrial parameters, but in science in general there are other parameters such as gravitational force, for instance. In modern fields of technology there are also other subtler parameters to be sensed, such as the recognition of patterns or of speech. We will mention these at the end of this chapter, as they are important in the emerging technology of robotics.

We will look at some of the common techniques of sensing the main industrial parameters above, with examples of their usage in industry.

## TEMPERATURE

Most sensor texts treat temperature in great detail as this was one of the first sensor technologies to be considered seriously. There are several classes of temperature sensor, and we will now look at those classes, and how they work. This parameter also gives an opportunity to show how sensors in general are interfaced to a micro system, and thus interfacing and associated techniques will be introduced as we consider temperature measurement. This will naturally lead onto the examination of sensors for the other parameters mentioned in the above list.

### Resistance measurement

Ohm's law states that the current flowing in a conductor is proportional to the applied voltage. Resistance is the constant of proportionality between the voltage and current. The formula is:

$$V = I \times R$$

V is voltage across the ends of the conductor, I is the current flowing in the conductor and R is the inherent electrical resistance of the conductor. R is meant to be a constant, but it is not, in general. It is normally dependent upon temperature, for instance. This variation in resistance with temperature can be harnessed to measure temperature itself, and is the basis of a low cost and easy to use temperature sensor. A precision resistance temperature sensor will be constructed to have a linear response to temperature variations, as well as an accurate and predictable resistance value for given temperatures.

To measure the temperature, therefore, a circuit such as that in Figure 3.1 might be used. In this case, a constant voltage, V, is applied to the

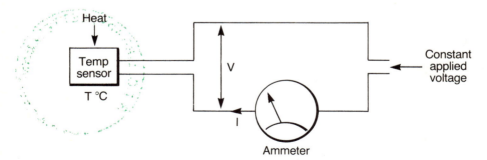

*Fig. 3.1* Basic temperature sensing.

temperature sensor through a current meter. As the temperature, T, changes, the resistance of the sensor changes, which affects I, and the reading on the ammeter changes to reflect the new temperature. If the temperature changes are large enough, and if the ammeter is sensitive enough, this will produce a possible, and very simple, method of estimating temperature. The meter needs to be converted, of course, as it will be in current units, while the reading is to be in degrees. Calibration is also essential in this setup as a number of variations to the ideal will be encountered, and they can all be allowed for by calibrating correctly.

The main problem with this setup is that the resistance variations are not great, and if a narrow range of temperature is to be sensed, the current variation across the scale will be small, and it will be difficult to measure the temperature accurately.

A better method of using the sensor is to increase its range and sensitivity by increasing the current variations using an amplifier. A further advantage can be accrued by including the temperature sensor in a resistance bridge. If this is unfamiliar, you will need to cast your mind back to ordinary school physics where resistance bridges are usually first encountered.

The essential point is that a circuit such as that shown in Figure 3.2 will produce a larger change of current into the amplifier, A, than that of the simple circuit in Figure 3.1. The mathematics of this setup are straightforward and will be found in any elementary electrical text book. The four resistances, of which the temperature sensor is one, balance one another out at some point, and the current flowing into the amplifier is zero. When one of the resistances changes, this causes a proportionately large change in the current through the amplifier, and thus great sensitivity can be gained.

The resistances in the bridge have to be accurate and stable, but the symmetry of the bridge circuit has a stabilising effect on temperature effects, for instance. This is an important circuit in the study of sensors in

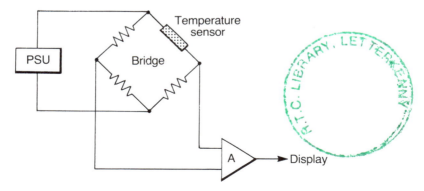

*Fig. 3.2* Practical temperature sensing.

general, and we will see it occurring again in strain gauges, pressure sensors, and anywhere else where electrical resistance is the basis of measurement. Indeed, we will see that the sensor manufacturers recognise the importance of resistance bridges and, in some devices, a resistance bridge is included within the sensor itself.

As this book is not an electronic treatise, it is beyond its scope to discuss the amplifier shown in Figure 3.2 in any detail. However, this is such an important and common element in control, that something must be said about it for completeness.

Such amplifiers are sometimes called 'instrumentation amplifiers', to denote that they are accurate and sensitive and will not interfere with the normal measurement variations of the sensors being used. The amplifiers used are often formed from one or more integrated circuits, each of which is called an 'operational amplifier'. The triangular symbol used in Figure 3.2 is the common representation of these devices. They normally have two inputs and one output, though many other lines may also be included on the device for special purposes.

Op-amps, as they are known, are general purpose amplifier elements which provide a set of standard and repeatable characteristics, as defined by their manufacturers. The integration of amplifiers for continuously variable signals is an important step in modern electronics; it saves the wasteful activity of soldering together many components onto a PCB or patch board in order to start using an amplifier in the real world. For a few pence, op-amps can now be bought which used to be formed from many separate components including, perhaps, expensive matched transistors. The saving in time is considerable, and the variation in op-amps offered by the many products on the market means that an amplifier for almost any need is at everyone's fingertips. This is particularly important in control electronics and sensors.

In use, you can think of an op-amp as sensing the imbalance of its two inputs. If they are at the same voltage, the output will be zero. If they are at even slightly different voltages, the output swings violently one way or the other, and most op-amps have gains of the order of millions if not kept under control. Even though Figure 3.2 does not show it specifically, there will be some other components surrounding the op-amp chip itself to keep it under control, and make the response predictable. These components will generally be assumed in the following, and the triangle symbol will be used to denote the complete amplifier circuit, including any required external components.

**Resistance temperature sensors**

There are several common types of resistance temperature sensor and two are particularly common – these are the platinum resistance sensor and the thermistor.

The platinum sensor is an example of a whole class of temperature sensors, made by utilising the natural variation of resistance of different materials with temperature. However, these all act in the same way, and the platinum version is probably the best known.

The thermistor uses a solid state technology, and its resistance changes with temperature can be quite large. Thus the sensitivity of measurement can be higher.

*Platinum resistance sensor*

Figure 3.3 shows a sketch of a platinum resistance sensor. A substrate has a thin film of platinum formed in a long track as shown on its surface. The substrate has to be rigid, and no strain must be put on it during normal use – you will see why when strain gauges are examined.

Connector pads are formed on the substrate to connect wires to the platinum track, and the whole sensor is normally sealed within a ceramic or other covering. The sensor can be quite small, a few millimeters on a side, and this helps to keep the thermal capacity small, thus allowing the sensor to react to temperature changes swiftly.

The temperature range of such devices depends much more on the substrate and covering than on the platinum, which has a high melting point. Typical ranges available will be from near absolute zero ($-273°$ C) to around seven or eight hundred degrees Celsius. The actual contact with the object being observed will depend upon the mechanics of the sensor. For instance, immersion in hot conductive liquids will require a fully insulating and fluid sealed enclosure, while a contact sensor, to be fixed onto a metal surface, will need a less robust construction where the sensing

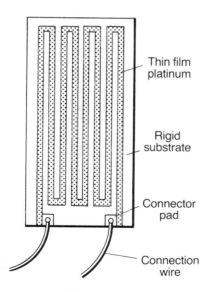

Thin film
platinum

Rigid
substrate

Connector
pad

Connection
wire

*Fig. 3.3* Platinum resistance sensor.

element is as little heat-insulated from the surface as possible.

Even though platinum is one of our most expensive metals, it is used in a very thin film form and hence sensors are cheap, and the simplicity of their interfacing helps to keep the total cost low.

Finally, there is even a world-wide standard of construction for these devices which defines the resistance at given temperatures. For instance, such a standard platinum sensor will have a resistance of 100 $\Omega$ at 0° C, and 138.5 $\Omega$ at 100° C. This also shows the degree of variation expected with temperature – 38.5 $\Omega$ per 100° C. This is not completely linear, of course, but its accuracy is comparatively good, and quite adequate for many industrial processes. To compare this sensor with the next one described, note that the resistance changes by around 0.385 % per °C, relative to its value at 0° C, and in a positive direction – that is, resistance increases as temperature increases.

*Thermistors*

Thermistors are fundamentally different in construction from the above. The sensing element is normally formed from a crystalline metal oxide and the resistance coefficient is most commonly negative – that is, its resistance decreases with temperature. However, thermistors can be made with positive coefficients. The resistance change can be quite high – as much as 5% per ° C, and as such they can be the most sensitive temperature sensors available. However, they are non-linear, and this has to be taken into account by the electronics using the sensor.

Thermistors can be fabricated as small beads and, as such, with their encapsulation, can be made with quite low thermal capacities. This means that their response times will be small. This could be important in a process where the temperature can change rapidly and the control must not lag too far behind.

The thermistor connects to similar circuitry to the platinum resistance sensor. Many different values of resistance and temperature ranges are available, and the manufacture of thermistors can be carefully controlled for a wide range of resistance values. The variation in resistance with temperature (temperature coefficient) is so great that it can be used directly, without amplification, in some applications. For instance, in a circuit which has to have a stable response over a given temperature range, a thermistor can be used to adjust the response as the temperature changes, to compensate for the whole circuit's natural tendency to change.

The thermistor is often used in non-microelectronic circuits such as in the temperature control of ovens and fridges. Again, the large temperature coefficient means that the device is straightforward and non-critical in use.

As with most temperature sensors, thermistors are available in a number of encapsulations from simple minimal ceramic coated beads to sealed metal probes and glass beads. The application determines the relevant encapsulation, and thought should always be given to protecting these small parts in any industrial process.

We will look at interfacing the two sensors into a micro controller later. Before that, however, we will examine another important temperature sensor, which figures in a large number of applications – namely the thermocouple.

### Thermocouple

In some ways, this is the most interesting and promising type of sensor discovered. It gives a direct method of converting heat energy into electrical current. The conversion around the other way is, of course, simple and used throughout our daily lives. However, the dream of converting available heat, and for that matter light, energy into other usable forms, via electricity, is still comparatively in its infancy. Its solution would allow us to dispense, for instance, with conventional steam production to turn generators in our power stations. Actually, we could dispense with the fuel burning too, in an ideal state.

In fact, the conversion of heat into electricity is deceptively straightforward. Simply weld a couple of dissimilar metal wires together, and the junction will automatically and ceaselessly produce electricity! Naturally, there are a few catches, and we will look at some of those now.

The first catch is that the quantity of electrical current produced is, by any measure, very small. Secondly, the first junction must be mated with another so that the two wires actually form a loop as shown in Figure 3.4. Only by maintaining the two junctions shown at different temperatures, $T_1$ and $T_2$, will a current, I, flow between the junctions $J_1$ and $J_2$. The current flowing depends upon the difference between $T_1$ and $T_2$, and on the two metals $M_1$ and $M_2$ chosen. This current is produced by the voltage

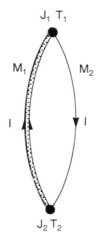

*Fig. 3.4* Thermocouple.

difference between the two metals in the junction, and it is this maintained voltage difference which is often given in the data on the devices. Only a very sensitive microammeter will detect the current under normal circumstances and it is, once again, necessary to amplify the result.

An advantage of this device is its minute thermal capacity – it can detect changes in temperature with great speed. Again, there are standards for the production of thermocouples and, again, they are encapsulated in various ways for different applications. As they can be made so small, they can often be inserted unobtrusively into spaces not normally suitable for a sensor. The main problems are their satisfactory adhesion to a surface and their protection against mechanical damage.

A practical thermocouple consists of two lengths of wire, welded together in an inert gas to prevent oxide from forming at the joint surface and distorting the electrical conduction. The second joint in the loop is normally emulated by the electronics which the thermocouple interfaces to.

One of the problems is to form a connection between the two wires of a thermocouple and the electronics, without forming yet more thermocouple junctions which will distort the output of the original one. Any two dissimilar metals, even twisted together, will form a thermocouple of some kind. The normal precaution is to extend the original pair only by using wire pairs exactly the same as the original ones, and to join them to the electronics using identical junctions, which will be maintained at the same temperature. Thus, as long as each thermocouple wire is, say, soldered to the electronics using the same type of solder, and to the same metal within the electronics, such as copper, any jointing effects will cancel out as long as the junctions are at the same temperature. This can normally be ensured by physically keeping these joints near to each other.

The standard for thermocouples is to give a voltage output when one of the junctions is at the measurement temperature and the other is at the melting point of pure water. This means that, theoretically, one of the junctions has to be maintained at the ice-point while the other is inserted into the machinery to be observed. This is somewhat impractical, and hence the need for a special 'ice-point compensation circuit' within the interface electronics. This is rather a complex affair, and it should come as no surprise to learn that the full electronics of a thermocouple interface and amplifier with compensator is produced by a number of manufacturers on a chip.

A typical circuit is illustrated in Figure 3.5. The chip mentioned there, the AD594 (or AD595), is manufactured by Analogue Devices, who are well known for their sensor conditioning circuits.

The thermocouple used here is one of two standardised types – type J or type K. Type J is formed from iron and constantan, while type K is chromel to alumel. These are standard metals and alloys, and the exact definitions of these materials determine the characteristics of the

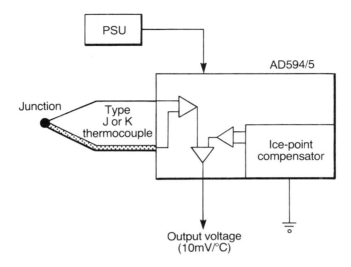

*Fig. 3.5* Thermocouple interface.

thermocouples. The AD594 works with type J, and the AD595 with type K

Within the AD594/5, there is an accurate and stable instrumentation amplifier which amplifies the output voltage of the thermocouple, and another which amplifies the output of the ice-point compensator which acts exactly like another junction immersed in melting ice. The two signals are then combined and amplified, and the output gives a voltage variation of 10 mV per degree change at the junction. This is linear and stable, and is a large enough variation to be used easily by cheap ordinary op-amps.

The advent of this type of 'monolithic' (i.e. solid state) circuit enables the previously rather problematic thermocouple to be used with complete simplicity. The circuit contains other internal modules too for such applications as checking that the thermocouple has not become open circuit, and for using the circuit as a set-point controller whereby a switch can be made when a given temperature is reached. Because the circuit is formed on a single chip, it is straightforward to allow for temperature compensation to minimise the effects of changes around the package itself – within reason. It also runs on a wide variation of power supply voltages, and is constructed to use a minimum of power for normal operation. All these characteristics are difficult to achieve using discrete components, and only in extreme circumstances, to produce special effects, would the classical discrete approach be used.

There are other chips in existence for a thermocouple interface, but the above is a good example of the type. The output can be used to drive an amplifier for showing the temperature on a display, or for reading into a micro controller. We will look at this latter shortly. As can be seen, the principle of measurement for a thermocouple is rather different from that of the variable resistance types of sensor, and thus the interface circuitry is not the same. However, you should note that the final output, a continuously varying current or voltage, is similar in each case. Not all sensor circuits give such an output, as we shall see.

### Other temperature sensors

Other temperature sensors might be found in more extreme circumstances, such as in the measurement of temperature in a hot furnace. In these circumstances, the best method of taking the temperature is often to use the radiation given out by the material within. This leads to a variety of devices generally called pyrometers.

A hand held optical pyrometer will normally contain a wire, electrically heated to glow red hot. The user looks through the instrument, with the furnace in the background, to compare the colour of the hot wire with that of the colour of the high temperature interior of the furnace. A control, calibrated in degrees, adjusts the current through the hot wire until the colours are identical. The temperature reading on the control gives an estimate of the furnace temperature.

An electronic sensor which uses the heat radiation emitted by a hot body will be similar to that sketched in Figure 3.6. The essential principles are straightforward. The incident radiation from the hot body is concentrated by an optical system onto a temperature sensor. This relates

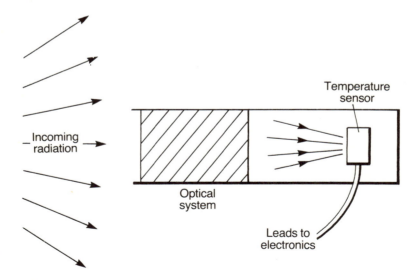

*Fig. 3.6* Pyrometer.

the characteristics of the radiation to the temperature of the emitter, and wires lead the electronic signal away to amplifiers in order to produce, once again, a continuously variable voltage or current.

The sensors used may be solid state detectors, or simply a set of thermocouples in series. Either way, this provides a useful non-contact method of measuring temperature, which can be important for some processes. We will see how this can be used shortly. An important point to note is that such sensors are not just used for measuring the temperature of red-hot bodies. Sensors of this sort can be manufactured for almost any range of temperature, but they are most common for temperatures well above ambient. They can be problematic in use, and design, due to the fact that every object, including the materials of which the sensor is constructed, will give off heat radiation to some degree.

Before we look at sensors for the other physical parameters mentioned above, it is important to explain how a micro system, which essentially uses binary numbers throughout, can interface with and read continuous voltages. Later, we will also see how it can produce such signals.

## INTERFACING ANALOGUE SIGNALS TO A MICRO

The technical term for continuously variable signals is 'analogue'. This is as opposed to the term 'digital' or 'discrete' which signals can only take on one of a number of set levels of value. Binary signals are only able to take on one of two levels – called '1' and '0'. Appendix 1 discusses the meaning of these terms if they are unfamiliar to you; 1 and 0 are simply convenient ways of describing signals which take on one of two possible states. In current micro and general electronic logic circuitry, the 1 level is given by a +5 V level, while 0 is given by 0 V. This is not completely universal, as we shall see when we look at data communications. In addition, there are logic integrated circuit families which work at other levels. However, we will be concerned with the most common types of circuit, and will assume these voltage levels unless otherwise stated.

To turn an analogue signal, from a temperature sensor for instance, into machine readable form, it must be transformed into a binary number consisting of these +5 and 0 voltage levels – i.e. into 1's and 0's. We will consider a black box approach to this problem, as the internal electronics are rather beyond our present scope.

Figure 3.7 shows an ideal conversion process. An analogue signal is input, which varies in some manner as shown. Each second, the analogue to digital converter (ADC) takes a reading of the input signal (V) and converts this to a binary number on its eight outputs as shown. The eight bits are labelled B0 to B7, with B0 being the least significant bit. At 0 seconds, V is at 0 V in this case, which corresponds to an output of all 0s from the binary lines. At the subsequent times, various binary numbers are

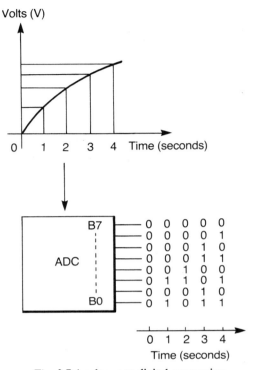

*Fig. 3.7* Analogue to digital conversion.

output to correspond with the input signals.

The main parameters of an ADC setup are the conversion times, the resolution of the output – dependent upon the number of bits given out – and the range of the input voltage. For instance, a 12-bit ADC might be used, with conversion time of 100 microseconds, and an input voltage range of 10 V. This means that the all 0s output corresponds to 0 V input, and all 1s output corresponds to 10 V. Furthermore, the output binary number will be within 100 microseconds of the current input value. The speed of conversion decides how much the reading may lag by in time. For a slowly varying, or DC, voltage a slow and low cost device might be sufficient. For analysing human speech in detail, however, the waveforms are complex and have sharply varying superimposed waveforms which require a high resolution and fairly fast ADC.

The resolution in the example above can be calculated as follows. A 12-bit number can have 4096 different possible values, from

0000   0000   0000

to

1111   1111   1111

We split the bits into fours to make them easier to read and understand.

All the possible binary numbers with 12 bits correspond to 4096 equally separated fractions of the range from 0 to 10 V. Thus, the 10 V range is split into 4096 parts, and each part is 1/4096 of the whole 10 V. This is the resolution of this particular ADC. It means that this ADC can measure a change of 0.00244 V in a 10 V range, or 2.44 mV. In comparison, an 8-bit ADC (1 in 256) will have an accuracy of 40 mV in a 10 V scale.

In a more general context, the resolution of an ADC is better stated by considering the number of bits used, and converting this to a decimal value. For instance, an 8-bit ADC has a resolution of 1 in 256, and this corresponds to an accuracy of $(1/256) \times 100\%$ or about 0.4%. A 12 bit ADC has an accuracy of 0.024% approximately. These resolutions are percentages of the full scale voltage applied to the ADC. If the input voltage falls between two of these voltage levels, the ADC makes a judgement as to the output binary number given, depending upon its construction.

To make this ADC read other voltage ranges, say from +3 V to +4.8 V, this voltage range must be fed through a voltage level converter to increase its range to 0 V to 10 V, unless the ADC has a control to scale its input voltage range automatically. The use of an ADC depends upon the accuracy of measurement required, as well as the actual voltages to be read.

Once the analogue voltage is converted into a binary number, it can be read by the micro in several ways. Some ADCs are LSI I/O chips just as described in the last chapter. That is, they have internal registers where the output binary numbers are held, and these may be read in the same way as any normal memory location.

Another method of input is to use some parallel input lines, as shown in

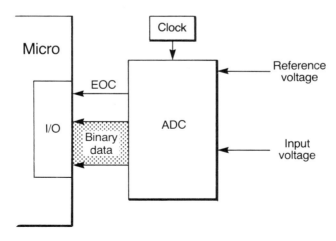

*Fig. 3.8* Reading an ADC.

Figure 3.8. This illustrates a practical ADC circuit. The micro has a set of input lines connected to the binary data from the ADC. In addition, a further control line is connected called EOC (end of conversion). The ADC itself has an input voltage to be measured, and a special reference voltage to scale the ADC's input range. The reference voltage determines the maximum voltage of the range, and any input voltage at or above this voltage will give an output binary number of all 1s. This often saves having to insert a voltage level changer between the input voltage and the ADC, and thus saves extra components, with their attendant addition to the temperature dependency and other inaccuracies of the system. The fewer extras inserted between the parameter being measured and the measurement device itself, the better.

The EOC line solves the problem of the micro trying to read a binary number from the ADC while it is changing its outputs, which it does at regular intervals. By observing the EOC line, the micro knows when the conversion is complete for any given cycle, and the output is stable and may be read. There may be many other control lines on the ADC, including one to allow the micro actually to command a conversion to begin, but these are beyond our present scope.

The ADC is shown being fed with a clock, which is used by the ADC to time its conversion cycles. A given ADC will be allowed a maximum clock frequency by the manufacturer, and the circuit designer may choose his own conversion rate, subject to this maximum frequency, by choosing the clock frequency.

In general, it is worth remembering that there is no point in specifying a 12-bit ADC for a parameter which only needs to be read to an accuracy of a few percent, nor an ADC which cycles at many kiloHertz if the parameter only changes on a time scale of a few seconds. It is important to design the ADC in for a specific purpose, and thus keep the total cost and complexity to a minimum consistent with performing the job in hand.

As an example of some design criteria, we will now look at some examples of the use of ADCs. Needless to say, they are used throughout process and product control, and it is important to appreciate their characteristics. Of course, as mentioned above in other contexts, the whole face of analogue parameter measurement has been changed over the past few years by almost universal use of monolithic circuits as ADCs. Gone are the days of constructing large modules of discrete circuitry on a board to produce an ADC. Chips now exist with many input channels, wide voltage ranges, and high speed conversion characteristics. In addition, they have the usual IC advantages of low temperature dependence, small drifts from the ideal, and extreme ease of use.

## AN EXAMPLE

As mentioned above, temperature is one of the most widely sensed parameters in the physical world. We will now look at an example which highlights both temperature measurement and the general design and choice of the ADC which goes with it.

The example is that of sensing the temperature of the plastic sheet within the thermoformer mentioned in Chapter 1. The technique uses a radiation pyrometer which can sense temperatures from around 30°C to around 300°C. As explained, it was necessary to use a non-contact sensor, and this is the only type found suitable.

The infra-red radiation from the surface of the plastic sheet has to be collected by the sensor without allowing reflection from the glowing heaters to interfere with the reading. Naturally, the heaters themselves are considerably hotter than the plastic sheet. This is a purely mechanical problem of mounting, and the pyrometer peers through a shaded hole in the lower heater bank at the plastic surface, and is positioned suitably to minimise reflection.

Pyrometer outputs vary between manufacturers, and we shall consider some typical characteristics so that the choice of ADC can be explained.

Outputs from sophisticated sensors will often be of the form of +1 to +5 V over the sensed range. Another standard is the 4 to 20 mA loop. In this latter case, two wires are used from the sensor to transmit the output current – a positive wire and a return. The use of a starting point at 4 mA instead of zero removes any uncertainty about the polarity of the signal around zero, and also ensures that there is always a signal being transmitted – this helps in failure detection.

We will assume that the pyrometer gives out a 4 to 20 mA signal linearly over the range from 30° to 400°C. Furthermore, we will assume the temperatures of interest to us here are from 60° to 250° only, and that the accuracy of measurement is to be around 1%.

We will see shortly that the linearity of the sensor makes the software and even the memory considerations of the controller simpler. However, a non-linear characteristic can often be catered for by the control system for a once-only development cost, and this may save a large amount of cost in these rather expensive pyrometers for each machine manufactured.

Firstly, the 4 to 20 mA signal (I) output by the pyrometer is converted into a voltage (V) by passing it through a stable and accurate resistance (R), which by Ohm's law will then have a voltage across it given by:

$$V = I \times R$$

There are more stable and accurate methods of current to voltage conversions using special op-amps, but we will assume that this is sufficient here. The required range of current output from the sensor is found by considering the temperature range required compared to the total range offered. Of the full range available, only 60° to 250° is needed. The first thing to do is to convert this into a current range. As the sensor is linear, this is achieved by simple proportion: 30° to 400° corresponds to a 16 mA swing from 4 to 20 mA. Thus, 60° to 250° corresponds to a range of:

$$16 \times (60 - 30)/(400 - 30) + 4 = 5.30 \text{ mA at } 60°$$

to

$$16 \times (250 - 30)/(400 - 30) + 4 = 13.51 \text{ mA at } 250°$$

If this were passed through a resistance of 1 K$\Omega$, the voltage variation would be 5.30 V to 13.51 V, which is a respectable range and will be used below.

Most ADC interfacing is eased by making the input voltage range start at 0 V. This could be achieved here by passing the voltages above through a level converter, but this should be avoided where possible, and is certainly not necessary here. For instance, if a reference voltage of 14 V were to be used on a typical ADC, then the full range of the ADC would extend from 0 V to 14 V for an output of all 0s to all 1s. In this case, therefore, with an input voltage of 5.30 to 13.51 V neither of these extremes would be met.

To obtain the accuracy of 1% over the range, an 8-bit ADC will suffice. This gives an accuracy of approximately 0.4% over the full range, and even with the applied voltages only occupying a part of the full range, we will see that the accuracy is sufficient. We will now calculate the actual temperature accuracy which will be gained by this setup.

Over a range of 0 to 14 V, the ADC resolves voltages of 1 in 256 or 0.055 V, approximately. In the case above, 5.30 to 13.51 V represents a range of 190°. This means that 0.055 V represents a temperature change of:

$$190 \times 0.055/(13.51 - 5.30) = 1.27° \text{ C}$$

Thus the accuracy is:

$$100 \times 1.27/190 = 0.67\%$$

This is an example of the type of calculation which is necessary to choose an ADC correctly. In fact, you should recalculate the above for an ADC having just 7 bits and an input range of 2 to 20 mA from the sensor. Assume that the reference voltage for the ADC is 12 V, but otherwise the same electronic setup is used. You should be able to see whether this would still be acceptable.

It should be clear that a useful object of ADC choice is to make the ADC's input voltage range as near as possible to the actual voltage variation measured. By this means the maximum resolution is obtained. If the input starts several volts from zero, instead of near to it, either an ADC with more bits will have to be used, or a voltage level changer will be required to bring the measured voltage down to a zero starting point.


## A SOFTWARE CONSIDERATION

Reading the ADC as connected in Figure 3.8, or in the case of an internally connected device, is simple. The software to perform this activity, even with the added activity of controlling the chip to perform conversions when required, and reading the EOC line is straightforward. Interrupts could even be used, as for the real time clock described in the last chapter. An interrupt could be generated each time the ADC finishes a conversion, and the interrupt routine would simply store the current value from the ADC in a memory location. Thus the main program would always have the current ADC output value stored in a 'magic' memory location without having to be aware of the process of reading the ADC through the I/O lines as required.

The problems arise when the voltages produced are non-linear. There are two main methods of overcoming this situation. The first is to apply a known given algorithm or mathematical formula to the data read in. Thus, if it were known that the sensor sticks strictly to a square law, the value could simply be square rooted, or something similar. However, this method is only applicable to sensors with known mathematical laws which are easy to 'decode'. The second method works for any sensor output, regular or not. It is the 'look-up table' method.

To convert the ADC's binary numbers into temperature, or any other parameter, each value read in is compared with a table of values held in ROM. No calculation is required, just a straightforward 'look up' in the table. A computer can be used to generate the look-up table from an empirical study, and the values can be blown into an EPROM, or formed onto ROM by some other method. The programming is simpler, but the development is more expensive and the memory requirements of the controller greater, as mentioned above.

An example of the use of a ROM-based look-up table in automatic control is in the compensation for irregularities in the movement of a closely controlled mechanical system. To machine material to produce the mechanics with extreme accuracy is costly. It would be much cheaper to machine to a lower degree of accuracy, measure the inaccuracies which remain, and program a ROM with this information in order to use it to correct for those inaccuracies.

The ROM could be interposed between the mechanical control and the item being controlled. Each value output by the controller could be turned into a digital value by an ADC, passed through the ROM, and thus changed in a manner which depends upon that value, and hence compensated and then converted back into an analogue signal to control the mechanics. This method is in fact used by a hard disk manufacturer to allow for rotational eccentricities in the disk while the magnetic head is moving across its surface under close control, to read or write a piece of data.

This shows how a look-up table in a ROM can be used as a 'transfer' function. The input data is transformed using the ROM data into some other form for processing.

## DIGITAL INTERFACING

We have looked at the interfacing of analogue signals to the micro system, but some sensors give out a digital signal of some form. For instance, a simple microswitch is either on or off, and hence its input to the micro is in one of two states. Figure 3.9 shows the circuit of a switch interface. Each input line is connected through a resistor to the +5 V line, and hence when the switch is open, the line is not left floating, it is held up at a 1 state. When a switch is closed, the line is held solidly at 0 V, which overcomes the 'pull up' resistor.

It is usual to arrange the electronics this way around as common electronic logic uses the zero state for its active value. This is largely due to the specific technology used to fabricate modern ICs. In general it is easier to hold an input line at a 1 value than a 0, and hence a resistor can hold a line high, but a direct connection to 0 V is needed to pull it fully low. In fact, a line left open or floating will normally float high, but this is not definite, and should not be relied upon. In fact, such a floating line can act as an antenna for electronic noise which will cause the input to jitter. This is a common source of noise problems within a micro system. You will see

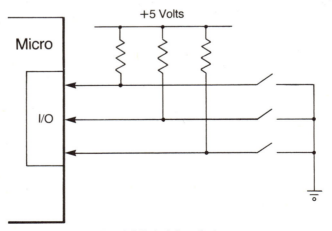

*Fig. 3.9* Switch interfacing.

pull up resistors used as shown in Figure 3.9 wherever there is a danger of a line being left to float.

Another form of digital interfacing is that already described for the ADC. A device puts out a binary number on a set of parallel lines, and these are connected directly to the parallel lines of a controller. Many types of sensor give this type of output directly, and we shall see some examples when we look at more sensors shortly. However, first we will look at another form of digital interfacing which is applicable to an important class of problem.

## Remote sensing and isolation

Consider the problem of sensing the temperatures within every machine on an extensive shop floor. Suppose the computer equipment is situated at one end of the building, and all the signals must be fed into this machine. If thermistors or thermocouples are used, analogue signals have to be transmitted long distances between the shop floor machines and the receiving controller. The problem is that any environment has a certain amount of electronic noise, but a shop floor of heavy machinery probably has more than most places. Small analogue signals will be interfered with easily and cause a distortion in the final reading. In addition, heavy machinery will have large AC voltages floating around them, and their earth connections, which form the return for any given signal path, will probably connect to a fairly 'dirty' mains earth. The delicate computer controller machinery must be isolated, ideally, from this environment.

Certain steps can be taken to limit this noise, but even using shielded wire suffers from the problem of high resistance between the transmitter and receiver, and shielding never fully eliminates the noise on the lines. The problem of isolation is solved using 'opto-isolators' or transformers. We will look at this shortly.

The best solution to electronic noise interference is to convert the analogue signals into digital signals. It is possible to send digital signals very long distances without degradation. The advantage of a binary signal is that it is either on or off, and there are no in-between values to be mixed up with varying noise spectra. Of course, it is still perfectly possible for electronic noise to interfere, and there is a whole branch of technology devoted to the sending and receipt of digital signals over long distances.

Assuming that a line can be constructed to carry digital signals without degradation, the question is how best to convert the analogue signal for sending. The ADC itself could be mounted at the sensor and its binary lines could be connected to transmission lines back to the controller. However, we have seen that eight or more bits are typical for an ADC, not counting the control lines which may be needed. On a shop floor containing a dozen machines, for instance, there would have to be upwards of 100 digital lines threading their way around the factory – a wasteful and expensive installation.

It is much better to try to condense the analogue signal into a single line of data. This can be done using an RS232 serial line, which will be explained in detail later in the book. However, this requires a fair degree of circuitry, and even intelligence, at the sensor end. An ideal solution would be to be able to use a simple and cheap circuit to convert the sensor output directly into a digital signal, and send it along a single line. There would then be a complementary circuit at the other end to recover the analogue data.

This can be achieved easily by a device called a 'voltage to frequency converter' or VFC. This is an electronic circuit which takes in the analogue signal and gives a continuously pulsing output whose frequency is accurately dependent upon the value of the input voltage. At the other end of the transmission line, these pulses are converted back into an analogue signal by a frequency to voltage converter (FVC). Normal analogue reading techniques are then used to input the signal's value to the micro system.

This is sketched in Figure 3.10, where isolation is also introduced. Here, the sensors at the machinery feed VFCs to convert the analogue signals into pulses. These pulses are fed along lines to 'opto-isolators', and then to FVCs and finally ADCs for reading into the central micro. The opto-isolators are devices for converting incoming electrical signals into light signals, and then back into electronic signals again. The advantage of these devices is that the two sides of the isolator are completely separate electrically. This is similar to radio transmission, where the electrical

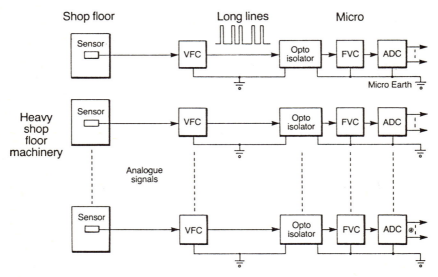

*Fig. 3.10* Remote sensing.

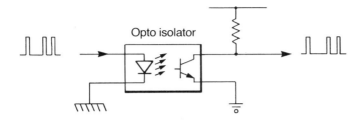

*Fig. 3.11* Opto-isolation.

environment of the transmitter is completely separate from that of the receiver, which may be a hand held transistor radio. This means that the earth of the heavy machinery is isolated from that of the micro, but the information carries through the system.

Another method of complete isolation is to use radio or fibre-optic communications through the system, and we will look at data transmission methods in a later chapter.

Figure 3.11 shows a diagram of a typical opto-isolator. The incoming data must carry enough power for the internal light emitting diode (LED) within the isolator. This LED simply converts the incoming waveform into light. The other side of the isolator may be a single light-sensitive transistor as shown, or a complete monolithic op-amp circuit for extra speed and sensitivity. The result is shown as a voltage developed by a load resistor, and will be with respect to the earth of the power supply shown at the top of the resistor – i.e. the micro's earth. As can be seen, short of high voltage sparking from one side to the other, the opto-isolator just passes the data without a direct electrical connection.

In a fibre-optic system, the distance between LED and light sensor is simply extended by passing the light along a light guide from the machinery to the micro. In that case, the isolation is effectively infinite, unless there are other paths within the factory for noise to pass along.

## FURTHER CONSIDERATIONS

When wiring a factory up in this manner, there are other considerations which are also important. For instance, electrical cables, by current safety regulations, have to be carried in special ducting, isolated from mains and other utilities. If fibre-optics is used, there are no special needs, and the overall system can be cheaper in the long run, despite the larger material cost of fibre-optics.

Another important consideration is the frequency of the signals being sent. Fast high frequency signals are more difficult to send and reproduce at the other end of the lines. Again, this is alleviated by fibre-optics which transmits almost without noise and distortion. The only serious effect suffered in fibre-optic transmission is attenuation along the fibre.

The work so far in this chapter has provided an introduction to the use and interfacing of sensors in general, and we have looked in detail at temperature sensing as an example. We will now turn to an examination of other types of sensor used in the industrial world. However, interfacing will be similar in many cases to the examples above.

## SIMPLE SWITCH BASED SENSORS

Many sensors just close a switch or send an on/off signal to denote some condition. For instance, we have looked at micro switches and how they are read. Any other type of switch will be the same. A float switch in a tank of fluid will close a contact when the level reaches a certain value. This is important in such applications as the automatic filling of a reservoir. A pump may be controlled by the float switch. A similar on/off switch is used in most domestic thermostats, from central heating boilers to refrigerators. Various mechanisms are harnessed to cause the parameter's value to close a switch. For instance, temperature expansion of a gas in a tube may be used to close a pressure switch when a given temperature is reached, and the heating agent is to be switched off. Such devices are self-sufficient and will work for many years day in and day out without fault.

Our interest, however, is to consider sensors which can be used to send back information about the real world to allow some process to react to the information, and control the system effectively. This is the basis of feedback.

The switch based sensors can be used for this purpose, as we have seen, and in many instances this is quite enough for the system's control. For instance, consider a controller for a large heating system, with electrical valves to each room, and room thermostats on every wall. The controller is only taking in switched inputs from each sensor, and if a given room is below its required temperature, the valve to that room is turned on by the system until the thermostat sends an 'on' signal to say that the room has reached temperature, and the valve can be turned off.

In fact, this system can probably be controlled by ordinary relay electronics, but the advantages of using a micro system are high. For instance, in a large installation, a full energy management system may save a very large amount of fuel overall. The system could be programmed to watch the temperatures throughout and, even when the system is supposed to be off, it could start the circulating pump to even up the temperatures throughout. Also, if the temperature drops outside to a certain level, the system could be programmed to come on a little earlier than usual, and so on. The whole of this system could function on purely on/off sensors, set in each room by the users of the installation. However, it would probably provide more flexible control if some analogue sensing were included.

## SENSORS OR TRANSDUCERS

The word 'sensor' means a device for measuring a physical quantity. The term 'transducer' is often used instead of 'sensor', although strictly, as its name implies, it is a device for changing one parameter into another.

At the start of this chapter, a whole list of parameters to be sensed was given. We will look at some of the more important ones now.

We have already seen temperature, and electrical current and voltage sensing via amplifiers and ADCs. Electrical parameters are the most easily sensed, and there are accurate volt and amp meters manufactured which not only give a readout on a display, but also provide a digital interface to a computer system. Needless to say this is also covered by standard complete chips for the purpose.

Common sensors in industrial applications measure strain (or force), position, weight, pressure, proximity and fluid flow. We will look at these sensors before passing on to other less often used devices.

### Stress and strain

In this context, it should be noted that stress is force per unit area and strain is change in length per unit length. These parameters are often locked up within a complex three-dimensional structure, and the sensors described below are applied to the surface of a body, and cannot fully measure the parameters within that body. Also, such sensors are only capable of measuring strain. The way in which this is related to the applied force is usually rather complex for any given body.

The commonest sensor for these parameters is the strain gauge. The principle behind this device is to cause small changes in resistance as a strain is caused by applied stress. Figure 3.12 shows the construction of a typical foil strain gauge. The principle of operation is as follows.

A small sheet of thin flexible plastic has a long conductive track formed on its surface, terminating as shown in connection pads. Thin flexible wire leads are soldered to these pads and taken to connector blocks, which allow more substantial wires to connect to the rest of the system. The strain gauge is glued carefully to the material to be observed with as little inherent stress as possible. As the material is stressed by an applied force, it suffers a strain and the strain gauge is stressed at the same time. As the strain gauge is light and flexible, it does not interfere with the strain of the surface to which the sensor is attached.

The strain in the conductive track causes its resistance to change in a predictable manner, and this is the basis for the strain measurement. This strain is related to the applied stress according to a complex relationship depending upon the physical nature of the body under stress. This is difficult to sort out theoretically, and the actual outputs from the strain gauge are usually calibrated by experiment for any given situation. Strain gauges do not give an absolute measure of stress directly.

These calibration data can be stored, for instance, within a ROM-based look-up table, and subsequently used by a controller to convert strain-gauge output into stress values, or applied force directly, depending upon the method of calibration. It is normal to use a computer system to take in the data from the gauge and the applied forces during calibration, and hence use the machine to generate the look-up data.

Typical track resistances of a foil strain gauge are of the order of 100 Ω, and resistance changes may only be a few tenths or hundredths of an ohm.

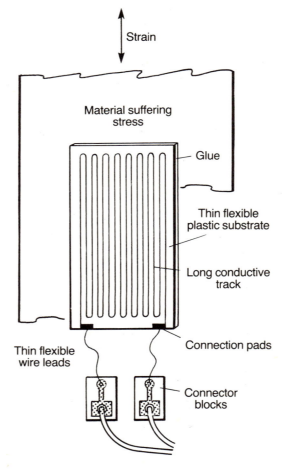

*Fig. 3.12* Foil strain gauge.

The strain gauge is treated in a similar manner to the platinum temperature gauge, and normally included in a bridge circuit for reading via an ADC by the control system. The strain direction is important and, as can be seen from Figure 3.12, will affect the strain gauge's resistance mostly along the direction of the double arrow shown. The object is to make the strain across the gauge have a minimum of effect. In this manner, two gauges mounted at 90° to each other can be used to determine the exact direction of any applied stress in the underlying material.

One of the problems in using strain gauges, or any resistive element, is the temperature dependence. A compensation for this can be achieved by adding another, unused, gauge into the circuit, and using them in a bridge arrangement. Figure 3.13 shows a possible setup. Two identical gauges $G_1$ and $G_2$ are placed as shown in the same arm of a bridge, with equal resistors R in the other. The proximity of the two gauges ensures that they are at the same temperature and hence suffer similar temperature effects. These effects can be shown to cancel out in the bridge circuit shown. The same, of course, applies to the R resistors.

The strain being measured is in the direction of the double arrow shown, and thus any variation in the output of the bridge will, theoretically, only be due to the strain in $G_1$, and not to any other effects. This depends upon the force being applied only along the arrow, and having no component

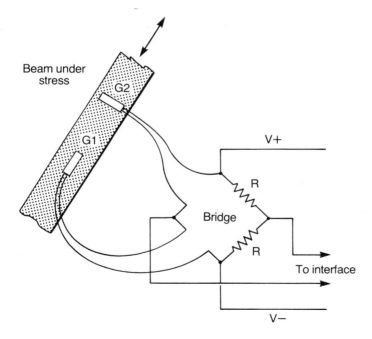

*Fig. 3.13* Temperature compensation.

across the beam. Any such component will affect the dummy strain gauge. However, in practice, a small effect will be noticed in the dummy, even with a strain purely across its width. These effects should be small, and in any case are compensated for by calibration.

Strain gauges are small and should have little effect on the measurement being made. This is ensured by making the contact with the stressed member as complete as possible, and choosing gauges with similar coefficients of thermal expansion to the material under observation.

Other applications of strain gauges include observations of strains in bridges and cranes, and in the construction of weight sensors. In weight sensors strain gauges are used within transducers called 'load cells', which are described next.

### Weight sensors

The measurement of weight is crucial in many process applications. It can be used to weigh items being produced and fed along a conveyer belt, or to ensure the correct fill for a container in a continuous production process. It can be used indirectly to measure the addition of a set of ingredients to a chemical process, and so on. Modern accurate weighing scales use load cells, and this ranges from weigh bridges in HGV checking, to the full gamut of the food processing industry. Many standard industrial processes require continuous or sporadic measurements of weight. This is even more so with current legislation which demands a close control on the stated weights and quantities of packed products.

Weight can, of course, be sensed by several different methods. For instance, a spring balance may be used, the extension or compression of which measures the weight applied and adjusts the value of a variable resistor through a mechanical coupling. A current passed through the variable resistor will cause a voltage drop across the resistor which is dependent upon weight applied. This voltage is then passed to an ADC. This suffers from various inaccuracies, and the fact that the load produces a large movement in the spring. This movement can often be a nuisance in a continuous weigher on a conveyer belt, for instance. Also, the spring balance can take some time to settle.

As mentioned above, the most important modern sensor for weight is the 'load cell'. This converts applied weight into a corresponding analogue voltage. Load cells are constructed, broadly speaking, in the manner shown in Figure 3.14. A frame of some kind is used to support a force, perhaps caused by a weight, and the strain in the frame is measured by the foil strain gauges shown. In this case, the frame is a piece of rectangular section metal beam, and four gauges are used. Using four gauges in this manner, and placing them in a bridge, will give excellent temperature compensation and good sensitivity. Only small variations in the strain are needed, and indeed it is important to ensure that the metal frame does not exceed its elastic limit.

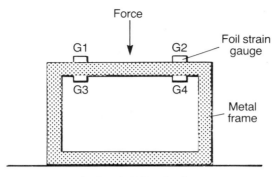

*Fig. 3.14* Load cell.

The variations in type and mechanical design of load cells are endless. Circular cross section beams may be used, either as a cylinder on end, or by applying the force across a diameter. Several load cells are often used in a weigher to save the problems of supporting a load on the top of a single small cell. However, the basic principle is the same throughout.

Other methods of measuring force are also found in load cells, including magnetic methods, solid state strain gauges, and a load cell based on an important method of position sensing called the 'LVDT'. We will look at this measurement element next.

### The LVDT position sensor

In some cases it is necessary to sense the exact position of a given part of a mechanical system at any instant. If it is simply necessary to ensure that it has arrived at a given position, a microswitch or proximity detector may be used. Also, the position of a given object can be measured by keeping track of its movements, and deducing its current position. However, the LVDT allows the absolute position of an object to be found, directly.

LVDT stands for linear variable differential transformer. Figure 3.15 sketches the principles of this device. At the top of the figure, three coils can be seen wound on a bobbin. The magnetic core is free to move through the bobbin's axis, and causes magnetic coupling from the central primary winding to the two secondary windings. The secondaries are carefully shown wound in the same direction on the bobbin, and cross-coupled together. Thus, when the core is at the centre of the primary, it couples the primary equally to the two secondaries, which produce the same anti-phase voltages, and thus cancel out to give zero output. Any movement from this position causes an imbalance, and the resulting output depends upon position, not movement, of the core.

The circuits needed to use this device will be familiar to anyone dealing in AC waveforms and the design of transformer elements. It will also be clear from such considerations that the result is linear and that the phase

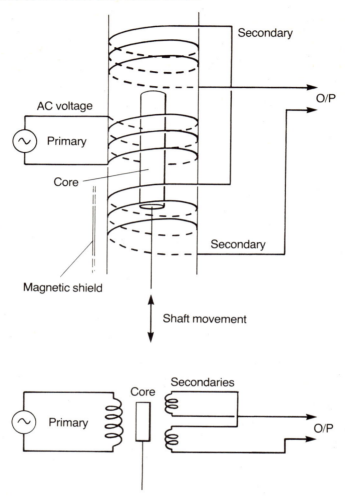

*Fig. 3.15* The LVDT.

difference between the primary and secondary waveforms can be used to
determine in which direction the core is displaced from the centre. It will
also be clear that using this device is a somewhat specialised activity, and
we shall not look at the electronics involved. Suffice to say that this device
gives accurate polarised position detection.

Advantages of the LVDT, apart from its linearity, revolve around its
mechanical characteristics. For instance, there is no frictional wear
involved, as the core does not touch the bobbin. Such devices are highly
robust and will last for many years. They can also be made to give quite
large output voltages for very small displacements, and the output is fully
analogue – there are no resolution problems, other than those caused by

the mechanical linkages and the associated electronics.

The LVDT can be used as the basis for a load cell in the same way as the strain gauge. A frame is used to allow a strain to be caused by an applied weight, and the small movements which are caused are mechanically linked to the core of an LVDT. The result is a voltage which is dependent upon the weight applied, and the mechanical setup.

Of course, once position can be detected, this can be turned into velocity and acceleration by a suitable processing of the time dependence of position of a large mass on the LVDT's shaft. This effectively makes use of inertia, and is the basis of the accelerometer, or acceleration sensor. However, velocities and acceleration can only be detected over short distances using an LVDT. It would be useful, for instance, in measuring vibrational movements, which are cyclic over short distances. For this purpose, a strain gauge would also be of use.

**Another sensor for angular and linear position**

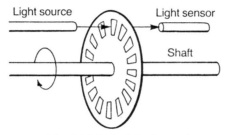

*Fig. 3.16* Optical shaft encoder.

A common sensor for position related parameters is the optical encoder, most commonly found in rotational elements. We will look at this device first, and then apply it to linear movement.

In crude terms, Figure 3.16 shows an optical shaft encoder, in order to demonstrate the basic principle of sensing rotation. A disc is fitted to the shaft, and rotates with it. The disc has alternately opaque and transparent bars arranged radially as shown. As the disc rotates, a light beam shining through the disc is interrupted by the bars, and a pulse train is generated by the light sensor. The micro system observes this pulse train, and by counting can tell the angle through which the shaft has turned.

A more accurate and high resolution disc system uses moiré patterns to tell small changes in angle. Such devices can also be used to detect linear movement, and this is sketched in Figure 3.17. Two gratings are arranged to allow a light source to shine through them as shown. The grating lines are at a slight angle to each other, and the characteristic moiré patterns are formed as they move with respect to each other. The dark patterns produce a pulse train as for the last sensor. These gratings are also used for angular movement sensing, by arranging one of them around the

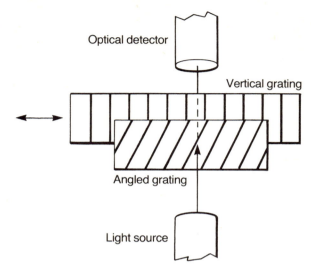

*Fig. 3.17* Moiré pattern linear movement sensor.

circumference of a moving disc, and the other as a fixed partial disc in front of the moving disc, with the light shining through both.

All the detectors described so far can be used for linear and angular position, velocity and acceleration. Velocity and acceleration are just a question of counting pulses and relating them to time by calculation within the micro system. This can also be performed by some simple digital electronics, to save processor time. In addition, a small design change to the basic mechanics of the sensors, and the addition of another optical detector, will allow the direction of the motion to be determined. However, absolute position is not possible with these detectors, and it is always necessary to home the mechanical system to some known position before starting. Alternatively, to keep the system    synchronised, a home position can be detected at regular intervals during the motion.

To detect absolute position, more light systems are needed, and a disc or track containing binary patterns is needed. Such a disc is shown in Figure 3.18. This has just three tracks and three light detectors. The three angular tracks are shaded and transparent in a binary pattern as shown, with the least significant bit nearer to the centre. As the disc rotates, the light detectors produce a 3-bit binary number depending upon which region of the disc in in front of the detectors. This means that the angular position of the shaft is known in an absolute manner.

Naturally, many more tracks than this are found on practical discs, but even then there is a fundamental problem with this particular pattern of disc. The problem is best seen at the top of the disc in the figure, where the all 1s state meets the all 0s state. If the disc should stop with this interface in front of the light detectors, it is impossible to predict what they will

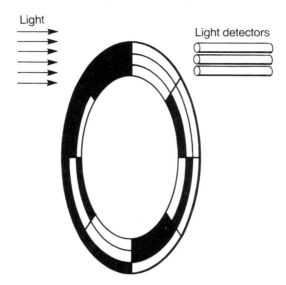

Light

Light detectors

*Fig. 3.18* Angle-coded disc.

register. This means that the electronics will probably be fooled into thinking that the disc is in some other position entirely, on some other part of the disc. This sort of indeterminacy will also occur, to a lesser degree, at some of the other interfaces.

This problem can be solved by several means. For instance, another track could be used to inform the electronics that this is not a valid position. However, a better way to solve the problem is to use a pattern such that each binary pattern is just one bit different from the preceding and following one. Such a pattern is called a Gray code, and this is the universal solution to the problem. This pattern will only be uncertain by one bit, and the uncertainty will always be between the two elements either side of the interface, and will never produce an output referring to a completely different part of the disc.

Just to show that this is possible, the following is the Gray code for 3 bits:

000   001   011   010   110   111   101   100

You can see that each pattern is just one bit different from the ones on either side, and the first and last also have this property, ensuring that whole pattern can be placed in a circle. It is then only a question of decoding the pattern back into binary, if this is required. It can be done conveniently by a single integrated circuit, or a look-up table within the micro, for instance.

Industrial applications for the precision moiré pattern position sensors include precision numerical controlled machinery. For instance, the X–Y table of an NC drill or milling machine will normally use these sensors to provide feedback to the controller of the table's postion.

An interesting application for these devices was found in a contact lens grinding machine. A small button of plastic was to be placed on the end of the rotating lathe shaft, and the surface ground down using a highly precise diamond-bit tool. The grinding was to be done to a depth accuracy of a few microns, with no slippage in the positional sensing allowed. This was solved using a moiré fringe linear position detector, and the whole head assembly moved back to a home position before starting. The total movement of the head before encountering the diamond tool was of the order of 30 or 40 cm, and then the position was to be measured by a few microns. The sensor was found to be precise and the whole system reliably counted the many pulses given out by the sensor during the major movement into place, and then accurately gave a feedback of position as the grinding took place.

Such sensors are also long-lived and highly accurate – this is an advantage of a digital sensor over an analogue sensor. There will be no temperature drift by the nature of the mechanical design.

Interfacing this type of sensor only requires a single input line, or two if the direction is also to be detected. However, in an example such as the contact lens grinder, the major movement of the head will involve a very fast counting sequence if the millions of counts are to be performed fast enough to allow the head to move at an acceptable speed. This will be a problem for most processors, and it was decided early on in this project to perform the counts by an external electronic counter. The electronic counter kept a current count of the position of the head, counting up or down depending upon the direction of the head. It could be zeroed when the head was driven to the home position, to allow the count to start from zero if any slippage had occurred, or when the machine was first switched on. The count was contained within several bits on the output of the counting cicruitry, and had to be fed to the parallel input lines on the controller to be read. By this means, the program had no work to do other than simply read the position directly when required, or zero the counter for the home position.

There are other means of sensing position and movement, and such specialist information will be found in any book on sensors.

**Pressure sensors**

Fluid pressure can be sensed by such simple devices as collapsible gas bags, or pistons in cylinders which trap a gas bubble, and mechanical linkages to movement sensors. It is also possible to read electronically columns of fluid in manometer tubes. However, there are two main sensors in general use, and these will be described below.

The first is the Bordon tube. This forms the basis of most of the direct

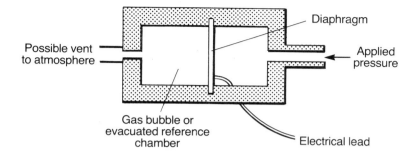

*Fig. 3.19* Diaphragm pressure transducer.

readout pressure meters in existence. The principle is to use the fact that when a pressure difference is maintained between the inside and the outside of a curved tube, the tube will tend to bend. This bending is then coupled to a movement sensor, and the rest is exactly as for any of the similar cases above. Sometimes a straight spiral tube is used, and the pressure difference makes the tube twist.

The Bordon tube gives a reliable but somewhat cumbersome method of reading pressures over almost any range, from small low pressures to the largest pressures encountered in any process or part of the ocean's depths. The Bordon tube is perhaps not best suited, however, to very fast moving vibrational pressures, as the mechanics takes a finite time to act. This is probably best covered by piezo-electric sensors with very low masses.

The second type of sensor is more important for electronic use, and is shown in principle in Figure 3.19. Its basic form is a diaphragm held between a reference chamber and an incoming fluid under pressure. The difference in pressure across the diaphragm causes it to distort, and this distortion is sensed by any one of the many methods available for strain measurement. For instance, modern gauges have strain gauges etched onto the surface of the diaphragm to allow the strain to be measured directly. In others, mechanical linkages within the reference chamber link the diaphragm to an LVDT, or a capacitive or other position sensor. Sometimes, a piezo-electric crystal is used to detect the strain.

These sensors are cheap, reliable and small. They are easily connected into a hydraulic or pneumatic line, for instance, and will react very fast to applied pressure changes. Their use is becoming universal in process and product control applications. If the pressure to be measured is to be compared with atmospheric pressure, then the reference chamber is vented to the air. Sometimes the reference chamber is evacuated to prevent temperature changes from affecting the reference pressure. The sensing can be performed over almost any pressure range, and they can be made to act with hot corrosive fluids with ease.

### Fluid flow sensing

Common flow sensors for low flow rates have a small light turbine within a transparent tube which is connected in series with a pipe holding the flow. A light beam is passed across the turbine, and as it turns with the flow, the beam is interrupted and a light detector produces a pulse for each pass of a turbine blade. This type of sensor has been used in the past for such applications as monitoring the flow of petrol to an automobile engine. It is reasonably accurate, and very low in cost.

The gas meters with which we are familiar in domestic residences use a different principle. The meter consists of a pair of bellows. Gas is allowed to enter one side of the bellows, and it pushes the bellows over one way and then the other, while dispensing gas to the consumer. The meter is powered by the pressure of the gas main itself, and the motion is transmitted to a gear train and indicator to show the amount of gas used.

Apart from these moving measurement methods, there are several classes of flow meter which have no moving parts, but which measure pressure along the pipe in one form or another. Figure 3.20 shows two such principles of measurement. The top example shows how a venturi in the flow pipe causes the pressure to change by an amount dependent upon the fluid flow velocity. The difference between $P_1$ and $P_2$ will give the flow rate. A baffle or orifice plate can be interposed in the flow instead, and

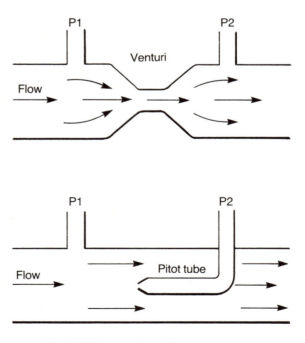

*Fig. 3.20* Pressure-based flow measurement.

again the differential pressure measured before and after the restriction.

The lower sketch in Figure 3.20 shows the use of a Pitot tube. This tube points upstream, and will show a positive pressure dependent upon the fluid flow. Again, the difference between $P_1$ and $P_2$ can also be used to give the flow velocity.

These pressures may not be high, and it is important to ensure that the method used does not allow fluid to leak out. Measurement may be by any of the usual methods, or a simple mechanical arrangement can link the tubes carrying $P_1$ and $P_2$ to either side of a sprung diaphragm or piston, and the displacement of the diaphragm will give the pressure differential.

The mechanical considerations for the use of flow meters are important in practical use. For instance, a Pitot tube cannot be used where the fluid contains particles of various sizes, due to its tendency to become blocked.

Another popular flow sensor, though more expensive, is the Doppler flowmeter. This uses the same principle which shifts the tone of a siren as it approaches and then leaves the observer. A high frequency (ultrasonic) sound wave is beamed into the moving liquid, and reflection occurs from any inconsistencies in the fluid flow, including turbulent vortices, particles, bubbles etc. The comparison between the reflected and transmitted signals gives a measure of the velocity of flow of the fluid. The great advantage of this method is that the pipe can be left intact, and the sensor simply attached to the outside surface. Unfortunately, the hardware is expensive and somewhat complex.

**Proximity sensors**

We have looked in principle at a wide range of sensors so far, and only a few important examples remain. The first is the proximity detector.

There are several types of proximity sensor depending upon the material whose proximity is to be sensed. A typical physical format of proximity sensor is shown in Figure 3.21. The electronics is confined to a metal cylinder whose body is threaded with a couple of nuts for a single-hole

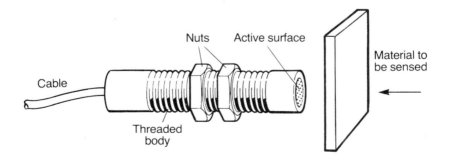

*Fig. 3.21* Proximity detector.

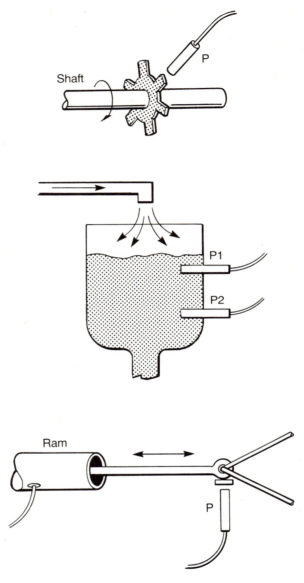

*Fig. 3.22 Applications of proximity detectors.*

fixing. One end of the detector has the active surface, and the other has a cable through which the signals and power pass. It is possible to buy detectors containing a complete electronic circuit which is powered by an external power supply, and which gives out a high or low signal depending upon the position of the object whose proximity is to be sensed.

The advantage of a proximity detector for noting the presence of an object is that no physical contact is necessary for the detection. The detector will normally detect the presence of the object within a few millimeters of the active surface. The general use of such detectors is in the detection of the end of a mechanical movement such as that caused, for instance, by a pneumatic ram. The proximity detector, in this application, supplants the microswitch which must be activated by a mechanical arm or lever.

The principles behind the design of the proximity detector range from inductive and magnetic to capacitive sensing mechanisms. The first two are used to detect the presence of metallic objects, while the third may be used for non-metals. This means that the presence of almost any material may be detected with some type of detector. However, the easiest material to detect is ferrous metal.

Figure 3.22 gives some sketches of typical uses of proximity detectors (P, $P_1$ and $P_2$). The top diagram shows the detector being used as a tachometer. A cog or cam on a shaft brings some material near to P as the rotation proceeds, and its speed or position can be measured as a result. Another application is in the detection of the high and low fill points in a hopper of material. The material may be fluid or a solid as shown. The proximity detectors can be mounted to be flush with the inside surface of the hopper. They normally come in a robust body, and can be made to resist corrosive materials. A common application is in the detection of the end of a mechanical movement. This may, for instance, be caused by a hydraulic or pneumatic ram, as shown in Figure 3.22. Either the natural mechanics of the setup is used to present a metallic surface to the detector, or a specially placed metal piece is welded onto a convenient place as shown.

A certain amount of selectivity can be gained by choosing the correct sensor and target for the job. Thus moving a length of plastic or wooden beam past an inductive sensor will not affect it, while a metal insert in the beam's surface will be detected when it reaches the sensor. Alternatively, an ordinary nut or bolt on the beam can be sensed without altering the beam's design at all.

## Other Sensors

The science of sensors is very wide and varied, and this book cannot possibly treat the field fully. For more complete and detailed descriptions of such devices, the reader is referred to specialised books on sensors mentioned in the Bibliography.

Some sensors rely on other laws of physics. An example of this is the Hall effect. In this effect, a solid state crystal has a current passing through it. An applied magnetic field produces a measurable voltage across the crystal. This effect can be used for proximity detectors where a mechanical movement brings a magnet up to the Hall effect transducer. This effect is somewhat more complex when used with AC magnetic fields. In this context, the Hall effect can be used to measure the difficult variable of AC power.

Almost any physical effect can be turned to use within a sensor, and more and more such devices are continuously being produced. Some of the parameters and sensors not mentioned so far include light, radiation and humidity. In addition, level sensors of various kinds have only been lightly touched on. Also not included are rotational transformer devices (synchros and resolvers), and so on. However, of these, light detection, which has been mentioned, is important.

Light detection is normally covered by one of three main types of sensor. These are photo-emitters, photo-resistors and semiconductor devices including photo-diodes and photo-transistors. The first type includes silicon and selenium cells which actually generate an electric current when light falls on their surfaces. Such devices are used to power calculators and satellites.

The second type of light cell simply adjusts its resistance with incident light. The change in resistance is normally very large, and will swamp any temperature effects. The last type of sensor is the most sensitive, and involves the change in characteristics of a solid state device. One of the most sensitive detectors in use is the PIN diode. The current through such devices is dependent upon light incidence, and they are often used as detectors in fibre-optic systems.

Optical detectors can be used to provide a measure of light level itself, but more commonly they are used to measure position or the presence of a given object. For instance, a lever or arm breaking a light beam can thus signal its presence. Alternatively, a reflective disc can be placed on the arm, and a combined light emitter and detector will detect the reflection, and hence the proximity of the object.

The industrial use of light detectors suffers from one serious problem — that of dust. Many processes are dusty and dirty, as are many apparently clean shop floors. It is important that the light emitters and sensors are kept dust free. This is also true, for instance, of the radiation pyrometer described for use in the thermoformer. In this case, a constant air stream is directed over the glass front window of the detector to prevent a build up of dust.

Visible light is not always used in industrial applications, in fact infra-red is rather more common. A low-light system will use an LED for the source, otherwise normal tungsten bulbs are used, for instance.

Figure 3.23 shows a common type of detector called a vane switch. As the shaft rotates, a vane breaks the beam from the emitter E to the detector D, and a signal is given out. Vane switches can be purchased containing all the electronics in a single module as shown, to give a sharp switched signal

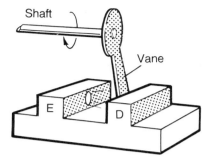

*Fig. 3.23* Vane Switch.

which will interface directly with electronic logic, without further interface.

Light beams can be used to detect and count the products passing along a conveyor belt, and can thus give automatic information to a data collection system.

### More exotic forms of sensor

In some applications, it is necessary to sense more complex data from the real world than has been mentioned so far. For instance, in a robotic system, it may be necessary to provide a rudimentary optical scanning system which can identify simple patterns. It may also be necessary to analyse and act upon special sounds, or even the human voice. The latter is still a major problem, and has yet to be solved. There are some simple but unreliable and restricted speech recognition systems in existence, but the general speech recogniser has yet to be produced.

We will look first at the problem of recognising patterns optically. This generally starts with a video camera. Applications of video recognition in process control include the recognition of special codes imprinted on various products. For instance, bar codes are now used universally in products for the supermarket shelf. Only recently have systems been introduced to scan these codes at the checkout to save the operator from having to key in each item. This takes the form of a scanning device, similar to a video camera, which looks up through a window in front of the operator. Each item is simply slid across the window at any angle with the bar code downwards, and the bar code is read. This allows the price to be retrieved from memory, and allows automatic updating of a stock control and buying system.

Another application of this type of technology uses a standard cheap video camera to record packed products being produced by a process. For instance, biscuits are produced and packed into cardboard boxes in a production hall. Several types of biscuit and sizes of container are produced from different parts of the hall, but they all pass out of a 'hole in the wall' on a conveyor into the packing section. The boxes are printed on all sides with a large bar-code type of pattern. As they pass into packing, a video camera scans them, and records each item according to its printed code. This gives instant access to the current state of finished product, and may be used to feed into some type of production system.

Naturally, this is best suited to a system where not too much diversity of packing is used, and where each item can be made to pass in front of the camera. The system is designed to allow the code to be read no matter what the orientation or position of the code or box passing the camera – including upside down.

Another application within a biscuit factory would be in the detection of broken biscuits. A controller connected to the video camera could be programmed with the pattern of a fully formed biscuit, within certain limits of variation. If a broken biscuit were to pass the camera, it would be recognised as such, and an air jet used to blow it off into a reject bin. This principle can, of course, be used throughout process industry to detect variations from the norm wherever needed.

There are two types of video camera which may be used in these applications. The first is the normal TV-type of camera where a flying spot of electrons within an evacuated camera tube traverses the optical field, and builds up an image on a special target. The result is an analogue signal corresponding to each line of scan across the light and dark areas of each line. Thus the picture is built up from between 300 and 600 analogue line signals. To read this data into a microcomputer system, the analogue signals must be digitised using a fast ADC, and thus converted into a set of binary values, corresponding to the light values along each line.

The second type of camera system is solid state. This comes in many different forms. All semiconducting devices are light sensitive to some degree, but they are usually confined to an opaque plastic or ceramic package. However, if the top is left off the package, or a window is formed in the top surface, this effect can be utilised by the right chips.

A typical configuration is the matrix image sensor based device. Here a set of light-sensitive solid state elements is connected in a matrix. The matrix may be scanned and the state of each element noted by some external electronics. These states can be stored in memory within a microcomputer and used to analyse the image formed on the matrix surface by a lens system.

The Snap camera by Micro-Products of Cambridge uses this principle. An image sensor with a window in its top surface is placed inside a small camera housing, with a bayonet fitting in the front for standard 110-format camera lenses. By this means, an image is formed on the chip's surface.

The image sensor used here is actually a memory chip – which is a matrix of solid state cells which can be switched on or off. Instead of scanning the target with an electron beam, the micro connected to the snap camera turns all the elements of the memory matrix on, and then reads the states of all the cells after a given time interval. Photoelectric effects cause the cells with the greatest incident light to be turned off within the time interval, while the others, in darker areas, remain on. By this means the micro builds up a representation of the image, which may then be processed and/or reproduced on the computer's video screen.

In the application mentioned above for detecting broken or deformed biscuits, Figure 3.24 shows the types of image which may be produced by the camera when shown on a computer video screen. $I_1$ would be a perfect biscuit and $I_2$ may be a typical broken biscuit; $I_3$, on the other hand, may be a deformed biscuit. As long as the camera is mounted above the conveyor belt along which finished biscuits flow, it will always be a constant distance from the biscuits. This means that a perfect biscuit, no matter what its orientation, will always have the same area. A biscuit which has a piece missing, as in $I_2$, will have a smaller area. It is easy to calculate area as this is simply the number of cells within the boundary of the image of a biscuit. By this means, $I_2$ can be distinguished from $I_1$, which is the ideal.

$I_3$ is a different problem. It is a distorted version of $I_1$, and may well have the same area, or as near as cannot be distinguished by this rather rudimentary camera system. The problem here is to detect that it is not a rectangle. This can also be done using a fairly straightforward algorithm, and this type of problem is at the root of pattern recognition.

The snap camera principle does not give the resolution of a TV camera, but it is much easier to interface as it is already digital. The system is also very cost effective. A system with more resolution will require much more processing, but would be expected to be able to check for correct biscuit shapes with great accuracy.

The basis of any such system would be a learning period first, where the controller learns the ideal biscuit shape, and also learns the degree of

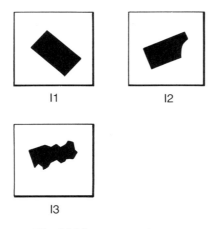

*Fig. 3.24* Snap camera images.

variation which is acceptable in the definition of a biscuit. This approach is also used in related pattern recognition problems such as those associated with speech processing.

Speech processing can be performed by digitising an incoming spoken word, storing it, and then using this data as a template for recognising the word later. However, a simple calculation shows how much memory and processing is really needed.

Suppose that one second of speech is to be digitised. Speech has a frequency band which extends to 4 or 5 KHz for sufficient resolution to be understood. This is for a quality on a level with the phone system. In order to use an ADC on the waveform from a microphone, the sampling, or conversion, rate must be well above this or it will affect the output significantly. It should be at least twice as fast. This means 10 kHz, or one sample every 100 microseconds. A second of speech, which may be two intelligible words, will thus require around 10K bytes of store, if an 8-bit ADC is used. It is possible for a microprocesor to take in data at this speed, but it is more efficient to use a separate piece of hardware to take in the speech data, and allow the micro to scan its memory for separate processing.

Greater speech quality simply requires more memory and a faster system. However, there are more intelligent ways to store speech, and there is a certain symmetry to the waveforms which we produce. This creates redundancy which can lead to more compact storage methods. A 10 to 1 reduction in the storage of speech data can be achieved, but this increases the complexity of the data decoding system which then has to replace all of the redundancy factored out of the original speech and recreate the original.

## EXPANDING THE CONTROLLER'S I/O LINES

It must be clear that in a complex process control system using many sensors, a large number of I/O lines will be required. A small controller may have trouble in catering for them all. This problem has been mentioned before, and requires some method of expanding the I/O lines to make a greater number available. This does not just involve input lines for sensors, it also affects outputs. Both can be expanded by the principle of 'multiplexing'. Chapter 5 gives a more technical description of this concept, but the following introduces the principle.

The idea of multiplexing is to allow many signals to be passed along just one line. The line is said to be multiplexed. A typical example of this is in the phone system where a single cable may be taking many conversations simultaneously. Another example is seen by recalling the principle of the data bus described in Chapter 2. There we saw how a group of lines was used for all data transfers to and from the MPU and external devices.

From one machine cycle to the next, the data bus may be used for input to the MPU from an I/O device, or output from the MPU to a memory location, and so on. This is an example of multiplexing, or multiple use of a set of lines – the data bus in this case. The MPU effectively controls the exact definition of the data bus's use from cycle to cycle. It is up to the MPU to ensure that no confusion arises from overlaps of data signals from instant to instant. This multiplexing means that just eight lines, in an 8-bit MPU system, are adequate for all the data transfer jobs throughout that system.

In general, multiplexing uses a given line differently from instant to instant – this is often called 'time slicing' or 'time sharing'. Both of these expressions are also used within the field of software to describe the multiplexing (effectively) of a given computer for a number of simultaneous tasks. The concept of multiplexing is fundamental to computing and control because all the processes occur sequentially, and they can only ever perform one type of operation at a given instant. In the future, the concept of parallel processing will become more and more common, and the effective power of computing devices will be increased by many times without the need for higher frequencies of processor clocks.

To allow a controller to read the state of many input signals using just one input line, the incoming signals must be switched onto that line by a selector switch, so that just one is switched through at any given instant. This can be done by hand by using a rotary switch. However, the controller can select the required signal more quickly using an external IC selector. The controller simply outputs a binary pattern to the selector and this commands it to let one of the signals through to the waiting controller input line. In this manner, the controller knows exactly which signal is being read, and it can cycle through the incoming signals at its own speed. We will see the electronics in a little more detail in Chapter 5.

# Chapter Four
# Controlling the Outside World

## INTRODUCTION

This chapter is concerned with describing some of the actuators which will be found useful in controlling products and processes. Actuators range from the control of mechanical actions such as motors and solenoids to non-mechanical actions such as the control of a bank of heaters or lights. The sophisticated and highly electronic end of control is in the display of data on a video screen, or other medium. At the other end are simple switches which turn a given parameter on or off – often implemented by relays.

We will look at some of the most important types of actuator, along with the most common forms of electronics which control them. At the same time, we will discuss their interface to a micro system. The idea of controlling the outside world also involves the tie-up between inputs and outputs – that is the concept of feedback. This will be mentioned below to show how control systems function.

## CLASSES OF ACTUATOR

Actuators come in many different forms. A common form is electromagnetic. An example is the simple electric motor which is turned on and off by a power switch. Alternatively, an electromagnet in the form of a cylinder can be made to eject a shaft placed along its axis. This solenoid actuator is good for short sharp powerful thrusts. Another important electromagnetic actuator is the stepper motor which can produce small rotational steps, in a digital manner. This rotational motion can be linked mechanically to linear motion, and thus small linear motion steps can be produced using a stepper. There are also other electromagnetic actuator systems including methods of linking two special motors together so that any movements of one are tracked accurately by the other. These are a little specialised, and will not be looked at here.

Another type of actuator is pneumatic or hydraulic with valves controlled by the electronics. Such systems usually consist of an electromechanical valve which controls fluid pressure to the actuator. The actuators can also be rotational.

In order to control the linear or angular position of the actuator, feedback from sensors on the actuator is applied to the electronic control. Such sensors can be microswitches or proximity detectors to sense when the movement is complete. Angular position can be detected by attaching a shaft encoder to a rotating hydraulic motor. Thus non-electromagnetic actuators can be controlled precisely by applying feedback techniques.

There are many non-mechanical devices to be controlled too, of course, such as heaters, lights, displays, and so on. These are usually quite simple to control as they are normally electrical or electronic in nature.

Before starting on the description of some typical actuators, we will look at the two most common electronic power switching techniques, along with their interface to the electronic systems. These are relays and solid state switches.

## POWER ELECTRONICS AND SAFETY

It is common in control systems to require the switching of high DC or AC voltages. Of these two, AC is the more usual, being the normal type of power supplied by electricity utilities. The reason for AC predominating is largely due to its ability to be voltage changed by simple coiled transformers. Typical mains voltages are stated as 110 or 240 V, but this is not a direct measure of the voltage which is supplied. AC voltages are continuously changing sine waves. Figure 4.1 shows the situation. The standard mains voltage in the UK is 240V, while 110 V is standard in the USA. The mains voltage, as shown, varies continuously, from zero to voltages greater than 240 (or 110). The frequency of the mains in the UK is 50 Hz, while 60 Hz is normal in the USA. This means that the sine wave completes its cycle in a fiftieth (or sixtieth) of a second.

In order to understand the voltages involved, consider an electric heater connected to the mains supply. At the beginning of the cycle it is turned off. As the cycle builds up, it heats up more and more strongly until it reaches the stated mains voltage (240 or 110 V). At this point it is heating at the rate which the stated mains voltage implies. However, it does not stop there. The voltage carries on rising and the heater heats up more and more strongly to make up for its slow start before reaching 240 (or 110) V. The same process is reversed on the downward slope of the voltage back to zero at half time in the cycle. Since the heater is not polarised in terms of voltage direction, the same heating process occurs in the negative half cycle. The result is that the amount of heat which is output over the changing voltages is equivalent to the heater being fed with a constant

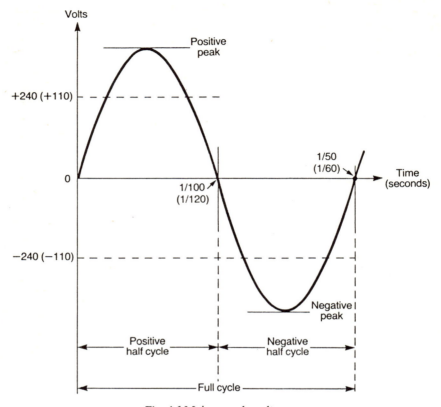

*Fig. 4.1* Mains supply voltage.

(DC) voltage of 240 (110) V throughout. We say that the AC waveform shown has a power equivalent of a 240 (110) V DC supply.

The DC equivalent is called the RMS (root mean square) value of the AC supply. It can be proven mathematically that for a sine wave as shown, the peak voltage would have to be:

(RMS value) × (square root of 2)

That is, for a 240 V supply, the peak voltage will be:

240 × 1.414 = 340 V approximately

and for a 110 V supply, it will be 156 V approximately. These voltages are supplied by the live side of the mains, and are with respect to the neutral, and usually the earth line too. This means that the live side of the supply, in the UK, traverses a voltage range of 680 V every fiftieth of a second. This is important in the connection of some types of circuit to the mains. This voltage is called the peak inverse voltage, or PIV, and capacitors connected in certain mains circuits will have this full voltage developed across them. It is not always sufficient, therefore, to choose components which are adequate according to the RMS value of a given mains supply. The actual voltages encountered can be nearly three times greater, and considerably more lethal.

Great care must be taken at all times when dealing with mains equipment, and it is important to understand the underlying principles of power supply as stated above. For instance, an innocuous looking capacitor can retain lethal voltages even after the supply has been turned off for some time. The lower voltages used in the USA will help to save lives under some conditions, but the voltages used are still dangerous, and can kill with ease. Remember, we do not have a sense which tells us of the presence of an electric field – any piece of metal in a power circuit could have a lethal voltage on it, though our eyes tell us that the component is safe.

Such voltages can even be induced without conductive connection to the power supply. A badly earthed chassis – a regrettably common occurrence in some process machinery – can hold a high induced voltage by being near to AC conductors. If you measure the voltage 'in the air' near power machinery, a high impedance meter will show that almost any piece of unearthed metal will contain a high induced AC voltage. In fact, we harness this type of induction in components such as transformers. It is true, of course, that the power behind small metal parts with induced voltages is low, but a large chassis can have a considerable ability to supply current when grounded through a conductor, including the human body.

Be wary and vigilant at all times when dealing with mains – there are too many lethal accidents each year. There are a few simple rules which should be observed. Firstly, do not be impressed by people who insist that they have had many 'belts' from the mains, and it does not affect them. Only the survivors of such incidents can tell this tale – many of their colleagues in industry die each year of instant heart failure due to electric shock. Secondly, while working with mains equipment, keep one hand in your pocket. This simple expedient can save you from sustaining a shock from one hand to another across the chest – where, you will remember, your heart is situated! You should also wear good thick shoes which provide insulation to earth – particularly in industrial sites where the floor will normally be concrete or metal and can provide a fairly good path to earth. However, you should also be aware that it is impossible to insulate fully against AC – alternating currents jump insulating gaps, which is, once again, one of the main properties used in AC circuits.

The simplest rule, of course, when dealing with power circuits, is to switch off whenever possible, and do not take chances with live equipment. You should also take the step of operating the power switch yourself, and do not rely on others to do it for you. You only have to make one mistake with such a communication.

## POWER SWITCHING BY RELAY

There are two main ways of switching high power from a micro; these are the relay and the TRIAC, or similar solid state devices. We will now look at these, along with some simple interfaces.

The relay requires little explanation – anyone working in engineering will be familiar with this humble and long established device. The principle is to use an electromagnet (solenoid) to pull a set of contacts open or closed when its coil is energised. Despite the introduction of a plethora of efficient solid state switching devices in modern electronics, and new ones being brought out continuously, the relay is still in common and general use in industry. At one time, there was a move to replace it completely, almost as an act of faith. However, the advantages of this electromagnetic component have always remained, particularly their ability to isolate the control electronics from that of the power circuitry. The move recently is to produce more efficient and neatly packaged relays for an industry which may never completely abandon them. Their disadvantages are that they are slower and somewhat less reliable than solid state circuits which have no moving parts or contacts to arc and wear out.

It is possible to buy relays which will handle low mains powers in packages which fit into the DIL sockets of the smaller sizes of integrated circuit. These DIL packages are neat to use, simple to replace and, as mentioned above, will provide almost complete isolation between the logic and the power. They are usually interfaced using circuits such as that shown in Figure 4.2. Here, a single output line is shown feeding a transistor interface to a relay. Even though this book is not concerned with exact electronic detail throughout, these circuits are so simple that they are worth explanation in order to provide some general background in the subject. It is also important for anyone contemplating a control system to have some idea of the number and complexity of components which will have to go into that system. It is useful, when considering a system, to know which components are simple, and which more involved.

The limiting resistor shown (typically 1 K Ohm) limits the current from the digital output line into the base of the transistor, to protect the digital system from overload. The transistor is used here as an electronic switch. When the output line is at a 0 level (0 V), the transistor is switched off. Thus the power supply, +V, is not connected through the relay coil to earth, and no current flows in the circuit. When the output line goes high under program control, the base of the transistor is taken to near +5 V, and this turns it on, as with a normal light switch. The circuit is thus completed through the relay, it switches on, and the relay contacts close or open in response. This gives a very simple method of controlling outside world devices from a program.

The flywheel diode shown is a protection for the transistor. Whenever

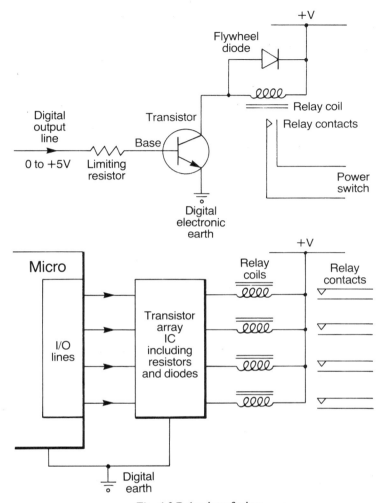

*Fig. 4.2* Relay interfacing.

an inductive (coil-based) load is switched, the voltage across the coil will develop a 'back EMF', which can be negative, even though the supply is positive, and may continue after the switching is complete – hence the term 'fly wheel'. This voltage could damage the transistor if allowed to feed back to it. The diode is a device which only conducts in one direction. When connected as shown, it is in the off condition. However, if a large negative voltage is developed across the relay coil, the diode will conduct and short this voltage out, thus protecting the transistor.

The circuit at the bottom of Figure 4.2 shows a typical use of several relays in a control system. It is normal to use an IC containing several transistor circuits as above, and hence save the need for many components to be soldered to a PCB when several relays are needed. It is important to consider the earth connection in any control system, as it is along such background paths that a lack of isolation may be discovered. The earth shown in these circuits is the normal 0 V line of the micro. The supply labelled +V may be the normal +5 V supply of the micro, or a separate larger supply for the relays. Either way, the power electronics is switched by the relay contacts, and remains completely separated from the digital electronics.

If higher power is to be switched than can be handled by these small relays, then they can be used to switch larger ones, including the usual types of high current contactors which are used on heavy machinery. The disadvantage of cascading relays in this manner is that extra power supplies may be necessary for each stage in the cascade. A solution is to use a power transistor circuit, instead of smaller relays, to amplify the low power digital outputs up to that required for switching the final power relays. The isolation is provided by these final relays.

## SOLID STATE POWER SWITCHING

The classical method of switching high power mains using solid state circuitry is to use the thyristor. This word is sometimes used to denote any type of power device as described below, and sometimes only for one of these classes. To save confusion, we will use other terms for each of the devices described.

We will look briefly at three devices and concentrate on only one. There are many other components which fall into the uses below, but these are best left to more specialised texts.

The three devices mentioned below are the SCR (silicon controlled rectifier), the TRIAC, which is a proprietary name now passed into common usage, and the GTO, or gate turn off device. The SCR was the first to be developed, followed by the TRIAC and finally the GTO, comparatively recently.

All three devices have three terminals, one of which is called the 'gate'. The gate is used in a similar manner to the transistor's base (see Figure 4.2). It controls the state of the power switching component. The SCR and GTO only allow current to flow in one direction, as with the diode, and hence their electronic symbols, shown in Figure 4.3, are similar to the diode. The anode is the positive terminal and the cathode negative. The TRIAC, on the other hand, can conduct in both directions. This means that if AC power is to be switched, every other cycle will be lost by the SCR and the GTO, as they can only conduct in one direction. The TRIAC, on the other hand, can conduct both half cycles of the mains, and as such can supply the full mains power. It is possible to connect several SCRs and GTOs in circuit to conduct each of the mains half cycles, and provide the same switching as the TRIAC, but the latter performs the action using a single component.

Another important characteristic of the SCR and TRIAC is that once the gate has been used to turn the device on, it remains on, no matter what the state of the gate, until the actual power supply being conducted through the device is turned off. The gate is thus sometimes referred to as a trigger, and the process is called an avalanche effect. This slight paradox means that to switch power on and off to a machine, it is necessary to have yet another power switch somewhere else on the power supply to switch the power switching device itself off after it has been triggered. This would be a serious problem if it were not for the fact that AC switches itself off 50 or 60 times per second anyway. This means that if a TRIAC, say, is triggered to supply AC mains to a machine, and its gate is then turned off, the TRIAC will remain on until the mains reaches zero volts next time – a maximum of half a cycle in time.

The fact that the TRIAC conducts both half cycles of the mains explains why it is the most widespread device for AC power switching. The GTO has the characteristic that it can be turned off by applying a reverse polarity voltage to the gate. However, as it only conducts in one direction,

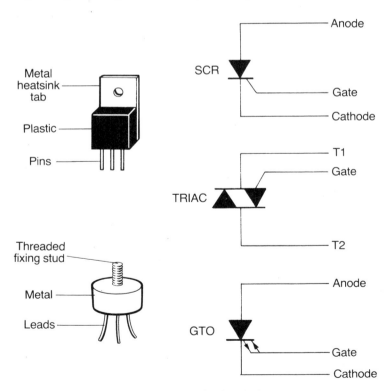

*Fig. 4.3* Power switching devices.

and the TRIAC turns off naturally anyway when conducting AC, it has not supplanted the TRIAC for AC power switching. It is, of course, of value in DC applications where the TRIAC is not successful.

Figure 4.3 shows some of the packaging which will be found for these devices. The first has a metal tab to which the solid state circuitry is thermally, and often electrically, connected to allow heat to be conducted away. None of these devices should be allowed to become too hot as their characteristics, including power handling, suffer as a result. The second package is a metal cylinder, which may be threaded, or have a threaded stud at one end for a single hole fixture to a heat sink. There are other packages too; some for very low power work have no heat sink connection, others have other forms of thermal connection.

The gates require a few volts, of a given polarity, for a few microseconds to trigger them. The TRIAC can actually accept a gate voltage of either polarity for triggering, once again showing how versatile this component is. However, there are certain gate polarities, depending upon the direction of current flow at that instant, which are more efficient for turning the TRIAC on. This leads to a little more complexity to some TRIAC circuits than might otherwise appear to be the case. However, you should be able to see that it is possible to trigger a non-critical TRIAC application from a

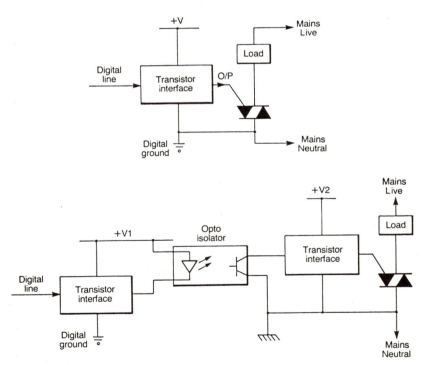

*Fig. 4.4* Non-critical TRIAC interfacing.

simple transistor circuit as for the relay. The disadvantage over the relay is that there is no isolation along the ground connection. This is explained by Figure 4.4.

The top diagram shows a simple transistor interface with a power supply +V, which may be the normal digital PSU. The interface is controlled by a digital line with respect to the digital electronics' own 0 V connection (ground). This must be fed to the interface as shown, and if +V is produced by another PSU, its 0 V line must also be connected to the digital ground. In addition, the output voltage from the transistor interface is with respect to this same ground, which must, therefore, be fed to the lower terminal of the TRIAC. As can be seen, this terminal will have to be connected to the neutral line of the mains in order to complete the circuit through the load being switched. This means that one side of the mains is in direct electrical contact with the digital circuitry.

The solution to this dangerous state of affairs is to place an opto-isolator in series with the circuit. Some TRIACs can be switched on directly by an opto-isolator, but in general it is more efficient to use another transistor interface as shown. This system completely isolates the power switching from the digital circuitry, and should be followed, at least in principle, on almost every occasion. The circuit is much more complex, of course, not least in its PSU requirements. Two completely separate PSUs are needed – $V_1$ and $V_2$. $V_1$ can often be the normal PSU of the micro to save one extra PSU. This circuit works as follows. The digital output sends a signal to the transistor interface which switches on the internal LED of the opto-isolator. This switches on the isolated output switch which switches on yet another transistor interface. This second interface is used for the TRIAC directly, and thus completes the control. To simplify this circuit, it is also possible to purchase opto-isolated TRIACs in a single package to save having separate components for this function, but they are generally of low power.

It should also be pointed out that although the above uses a DC trigger voltage for the TRIAC, it is more efficient, and sometimes essential, to switch the gate of a TRIAC with an AC signal of the same frequency as the power being switched. In this case, a low power TRIAC (perhaps opto-isolated for circuit simplicity) can be used to switch a mains-derived AC signal onto the trigger of the final TRIAC. This was how the problem of controlling the heater banks was finally solved in the plastic thermoforming machine mentioned in the previous chapters. Here, hundreds of digital lines had to control the heater TRIACs on a cycle by cycle basis with complete precision.

Another important consideration in some installations is the problem of causing electrical noise by the switching process. If a TRIAC is switched on after the AC mains has passed its zero point, then the load is suddenly subject to a voltage as high as 340 V in the UK. This sudden surge of power, quite different from the smooth sine wave build-up from zero suffered in each complete cycle causes considerable electromagnetic interference. To experience this directly, try holding a radio up to a domestic light dimmer, which normally switches the TRIAC after the mains zero-crossing point. The electrical noise given off is quite high. However, it is important to note that when the mains waveform returns to zero, at the end of each half cycle, the TRIAC is naturally switched off, without a sudden jump in power, and thus the switch-off point of the TRIAC is without electrical noise. Only the method of switching on needs attention.

The solution to the switch-on problem is to ensure that the gate is only triggered at or just after the start of a cycle, and before the voltage has built up significantly. This is called 'zero-crossing point' switching. A controller can monitor the mains supply and look for the zero-crossing points and ensure that the digital lines are only switched just after them. This was used in the thermoformer to save the hundreds of kilowatts of heaters being switched at any other point in the cycle. The result would have been too noisy to be tolerated by most customers for the machines. Thus, each half cycle of the mains was monitored, and all the heaters which were to be switched on for any given cycle were all switched at the start of that cycle. By adjusting which heaters were to be on, and for how many half cycles per second, the heat output of each heater was controlled to ensure that its duty cycle was correct for the particular part of the plastic sheet over which it was most effective.

The heater control above is an example of TRIAC 'burst control'. That is, a controlled number of mains half cycles are fed to the heaters each second, and by this means the duty cycle of the heat output can be varied. Another important control method is 'phase-angle' control. In this method, the electronics detects the zero-crossing point of the mains, waits a controlled time and then triggers the TRIAC after the zero-crossing point. The TRIAC then switches off naturally at the next zero-crossing point. In this manner, only part of each half is fed through. By controlling the amount of each half cycle which is passed through in this manner, the amount of mains power is adjusted. No power is lost, and the control is quite efficient. However, as mentioned above, the sudden switching on of mains power part way through its cycle creates electronic noise.

As has been seen, the TRIAC can be used as a power relay, but it must be isolated from the digital controlling electronics. The only systems where this isolation is not necessary is when the controlling electronics is naturally isolated from the rest of the system. For instance, in a light dimmer no opto-isolation is used as the whole of the control electronics is confined to a completely insulated enclosure. It is possible to do the same for a small single-chip control system, but in general this is rather difficult to achieve, and isolation will be required for safety.

## FEEDBACK

A general control system will consist of a method of controlling one or more parameters, and some sensors to observe the effect. The sensors tell the controller when the parameter has been changed sufficiently. This is called 'closed loop' control. Some control processes do not need feedback, and this is called 'open loop' control.

An example of closed loop control has already been given. In the thermoformer, the plastic sheet is heated until the pyrometer tells the controller that the required temperature has been reached on a representative part of the sheet. The heaters are turned off at that time, and the sheet indexed onto the forming station.

Our bodies are vast complexes of closed loop controllers – every time we lift up a cup without throwing it at the ceiling, or pat a small child on the head without knocking him out, we are unconsciously using closed loop control to apply just the right amount of strength for the job, and no more. In fact, the situation is considerably more complex than that. There is rarely a single parameter such as strength to be controlled in a human movement.

An example of an open loop controller would be a straightforward sequencer which did not check on the outcome of its actions. For instance, in a simple washing machine controller, there are parts of the cycle which are open loop. The object of a washing machine is to wash clothes, etc. However, no feedback is taken from the washing drum to see whether there are clothes present. The sequence will continue through wash, rinse, spin dry etc., regardless of the presence of any washing, and then regardless of how clean or wet or dry it may be. Each part of the cycle is of sufficient length to wash most things clean and comparatively dry by the time the cycle is over. Not every type of dirt will be removed, but the controller is designed to wash the majority correctly. This means that it will be too long for some things, and not long enough for others. This is typical of open loop control – the designer has to experiment, and finally guess the best operating characteristics, and take a compromise.

In this case, it would be rather difficult to design the machine any other way – how do you assess how dirty a given garment is? Also, a washing machine does usually contain some closed loop control, such as in the heater section. A thermostat allows the heater to be switched on just long enough to reach the correct temperature. If hot water is piped in, the heater is only used for a short time, while cold water causes the heater to heat for longer. This is a good example of closed loop control, and temperature is probably the most common feedback parameter in use.

We will see how feedback is used in control as we examine some more actuators, and their applications.

## ROTATIONAL ACTUATORS

There are many types of electric motor, and they can all be used for simple on/off movements, or for more complex variable position and speed modes of actuation. However, most of them require feedback of some kind to be controlled effectively. To see an example of a common form of control using motors, we will look at an industrial example of a special kind of actuator – this is a linear actuator, and is sketched in Figure 4.5.

The mechanics consist of a long threaded shaft which can be turned by a DC motor. On the shaft is a threaded boss with a fixing lug which may be bolted to anything which is to be moved by the actuator. The boss is prevented from spinning by the housing of the actuator, which is not shown. The relationship between the motor's revolution and the boss's linear movement is rigidly fixed by this arrangement, and hence the angular movement of the shaft exactly measures the linear movement of the boss.

For feedback of the boss's position, therefore, an optical shaft encoder is used. It does not need to be directional as the controller will know in which direction it has commanded the DC motor to turn. One advantage of the DC motor is that it is easy to control its direction – simply change the polarity of the supply voltage.

Another method of checking on the boss's movement might be considered to be the electrical power supplied to the motor – how much current for how much time. However, this is an example of a feedback parameter which is rather removed from the actuation itself. It will be dependent on such unpredictable effects as friction and load. It is important to observe a feedback parameter which is rigidly connected to the final movement, and preferably giving digital output such as the shaft encoder chosen above. The number of pulses from this device gives a precise and unchangeable measure of the distance travelled.

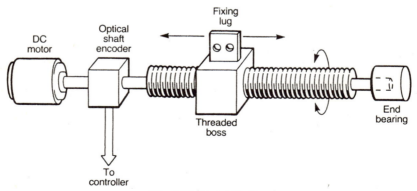

*Fig. 4.5* Linear actuator.

This produces a new problem, which is very common in control situations, and has been mentioned before. How does the controller know where the boss is at any time? It only knows how far it has travelled, and it does not even know this when it is first turned on, unless the data has been specially stored in non-volatile memory.

It is possible to use an absolute feedback parameter which gives the exact position directly at all times, but this is cumbersome and it is often difficult to gain sufficient resolution. If absolute position is required at all times, it is better to arrange this using some electronics which will retain its memory when switched off.

The more usual method of controlling this type of actuator is to place a microswitch or other detector at one end of the shaft, to sense when the boss is standing at some home position. The action the controller takes when switched on is to rotate the shaft to bring the boss to this home position. At that point, the count of pulses from the encoder is zeroed, and the control can commence. It is also good policy to home the actuator as often as possible during the control cycle to correct any inaccuracies in the position sensing which may have occurred. From time to time, a pulse may have been missed, particularly in turning around from one direction to the other.

There are two approaches to the collection of pulses from the encoder.

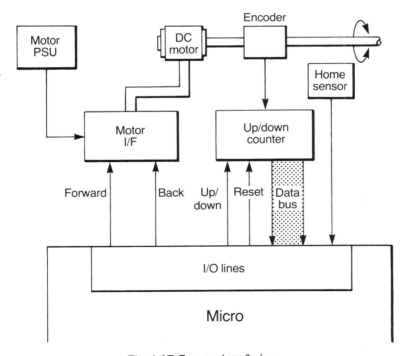

*Fig. 4.6* DC motor interfacing.

Either the controller itself has to count them, or some external electronics, an up/down counter for instance, holds the current count, and can be read by the controller whenever necessary. This latter solution is more practical in a system where the resolution of measurement is sufficiently high to necessitate very fast counting while the motor is turning.

The control system shown in Figure 4.6 gives some idea of the amount of control equipment needed for a DC motor controller, which just controls the position of the movement. No control over the speed of the motor is included here. In general the speed of the motor will also have to be controlled, and ramped up and down smoothly to save any overshoot on the required position.

The controller is able to control the motor's direction of travel by activating back or forward control lines into the motor interface (I/F) which contains power circuitry to the motor's windings. The controller also signals to the up/down counter the direction in which the motor is turning. This respectively increases or decreases the internal count on each pulse received from the encoder. At any instant, therefore, the controller can read the position of the boss from the value held in the counter.

In order to start the count from a known position when first switched on, the controller sends the motor back until the home sensor signals. At this point, the controller switches off the motor and resets the counter to zero. The home sensor must be chosen to be quite accurate in its position sensing as the whole position control proceeds from this starting datum. Microswitches can give an accuracy of a thousandth of an inch, while optical sensors will be more reliable, and can give similar accuracies.

It is also possible for much of the control mentioned above to occur automatically without the controller being involved. For instance, the home sensor can be connected directly to the reset of the counter, and to a cut-out on the motor's backward direction. Similarly, the up/down control of the counter can be connected directly, with a little electronic logic, to the direction lines of the motor, and thus change direction of the count automatically. This saves some I/O lines. However, for precise control of all these parameters, along with ramping control to allow the motor to arrive accurately at its destination without overshoot, a complete control system as shown will be needed.

This example gives some idea of the types of control used on DC motors. Needless to say, whether they are connected to a linear actuator, a robot arm joint, or a crown wheel and pinion in a model motor car, the control philosophy is the same.

Other types of continuously rotating motors are also used in some circumstances, but the DC motor is the most common in closed loop control. In fact, it is possible to purchase standard controller boards which require a minimum of control interface to give high level commands to the motor without becoming involved in the actual control itself. Such controllers normally have an inbuilt MPU-based controller purely to handle the motor.

The other type of control motor in general use is the stepper motor, and we will examine this now.

### Stepper motor

The DC motor suffered from the problem of requiring a fairly complex and expensive controller including full closed-loop feedback. For full and accurate control, it is even necessary to regulate the speed with a certain degree of intelligence. The problem is that a DC motor will simply spin when a voltage is maintained across its windings. The ideal type of motor would only rotate by a single accurate step for each pulse sent to it. In this manner, digital output lines could control it directly, via power amplifiers. This is precisely how the stepper motor works.

The internal components of a stepper are fairly simple, but we are more concerned here with the applications and design criteria associated with the device. The details can be found in the references in the Bibliography. Suffice it to say that a series of solenoids is used to drag a magnetised armature around. The connections to a stepper vary according to the exact design, but there will usually be at least four coil leads to the device plus a common connection, and by controlling the voltages on these leads, the stepper's armature can be made to rotate through specific small angular steps. It is possible to control a stepper using four output lines from a controller, via power amplifiers. However, this is a little wasteful in terms of the software and control time required, and it is usual to use one of the excellent low cost stepper motor control chips. Figure 4.7 shows a typical setup.

The stepper shown has four windings, and by controlling which windings are energised at any time, the shaft can be dragged round in one direction or the other by a single standard angular step. The standard angle stepped is entirely dependent upon the mechanics of the stepper, and may be typically a little over a degree. However, by controlling the windings, it is also possible to force the stepper shaft to rotate at half this standard step value – hence the full/half step line from the micro. The disadvantage of half step is that in exchange for the greater resolution of mechanical control, the torque generated by the motor in half step mode is rather less than in full step mode.

The controller chip accepts signals from the micro to select the step mode (full or half), and the direction of the rotation. It then receives pulses along the step line, and turns these into the necessary controls for the stepper windings, in order to step the motor once for each step pulse received.

The stepper itself may be connected directly to some of the chips on the market, as shown. Otherwise a set of power transistors can be used to

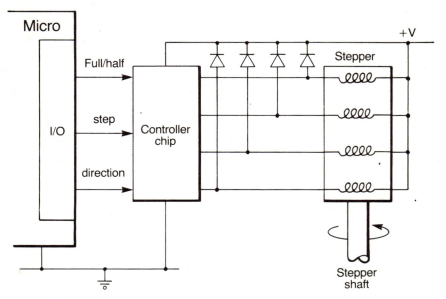

*Fig. 4.7* Stepper motor control.

convert the logic outputs of the controller chip into power pulses on the windings of the motor. The windings are shown here with the usual flywheel diodes as described in the last chapter to prevent negative voltage spikes from destroying the output transistors. A separate power supply (+V) is shown being fed to the stepper. In larger applications, quite high currents are required by the stepper, and design of the power supply is always an important step in stepper system design. Typical voltages required will be +24 V. Unless a set of opto-isolators is used in series with the control signals, the stepper earth will be connected to that of the digital system, which normally presents little problem as long as the normal decoupling, or smoothing, components are used in the digital power supply.

The stepper is so simple to interface, and use, that it is preferable to the DC motor wherever possible. However, there are some problems, and we will look at these now.

Since the angle of the shaft is so finely controllable, it is often not necessary to have a shaft encoder to feed back position. The micro simply keeps its own internal count of the steps. However, under excessive load the stepper motor can be made to slip and lose pulses. This causes the micro's count to be incorrect. Also, as before, a home position sensor is normally needed so that the count can be started from the home position when the system is switched on. Thus, by homing the stepper from time to time, or observing the home sensor on each revolution if possible, the count can be checked against reality, and adjusted if wrong. It is important to perform a fairly comprehensive load calculation during the choice of stepper for a given application. Furthermore, it is not possible to drive steppers with the angular speed of a DC motor. This means that they are used for applications where high speed is not needed.

A certain amount of compensation for these shortcomings can be achieved by carefully accelerating the speed of the stepping up and back down again from position to position. This saves some slipping, but requires rather more sophisticated control than for the above. This 'ramping' can be performed by the software within the controller which has complete control over the frequency of the steps. Once again, there are automatic controllers on the market which will accept data of the destination position of the stepper's shaft, and perform the complete control of the number of steps, including ramping to achieve the movement. Such controllers are an expensive alternative to software, but if time is tight within the processor, they may be necessary.

*Applications*
Steppers are used in a wide variety of product and process control applications. For instance, electronic watches often have sub-miniature steppers to move the hands. The new revolution in robotics has utilised the stepper to the full. Industry uses such devices for simple actions such as pick and place applications, automatic welding, and some heavy engineering.

Figure 4.8 shows a sketch of a robotic arm which can be used as an output peripheral for a micro controller. In this case, each joint and degree of freedom is shown being actuated by a stepper motor – five in all. This

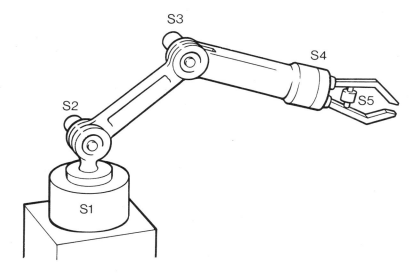

*Fig. 4.8* Robot Arm.

includes rotation of the whole arm, S1, two arm joints, S2 and S3, rotation of the gripper, S4, and gripper closure, S5.

The control of this system would be very precise, but there are one or two mechanical problems which do not show themselves at first sight. Firstly, to give the arm any real power to lift more than a few hundred grammes, the steppers would have to be fairly powerful. This is no problem for S1 and S2, which can be made as large as required. The problem comes with the actual masses of the rest of the control gear. The other steppers are at the end of a long moment arm, and cannot be too large. There is a trade-off here between power and weight — as in many other contexts. For an application where very small light objects have to be picked up from a supply bowl and placed within a structure with some precision, there should be little problem. The problem comes with weights of the order of a kilogramme or more.

One solution taken by most robot arm manufacturers is not to position the steppers on the arm as shown, but rather to transmit the action by toothed belts or wires. The steppers are positioned around the base of the arm and work it from there. Other compromises include no rotational control of the gripper, and perhaps just a solenoid to open and close the gripper rather than control it precisely. This all helps to reduce the weight of the assembly, and leaves more lifting capacity for the target load.

Other applications for steppers include the head movement in low and medium speed magnetic disk drives. As the magnetic surface of a disk rotates, data is stored on annular tracks by moving the head to a precise radial position, and reading or writing magnetic information on the surface. Steppers provide the precision of movement, when converted into linear motion, for this type of task. A stepper could have been used in the linear actuator described earlier in this chapter. The control would have been rather simpler, but the speed and torque are easier to achieve using a DC motor.

A class of steppers used to provide linear motion has a threaded hole through the central magnetic armature. A threaded stud can be screwed in and out of the motor by stepping the armature round. The threaded stud must be prevented from turning by the mechanical setup of which it is a part.


## ANALOGUE OUTPUT

So far, we have looked at output from the control system which has been entirely through digital lines. Often, an analogue output is needed to control a parameter which is continuously variable — for example, the sound level from an amplifier, the light level in a room, the iris in a camera lens. These are all analogue control parameters, and often require direct continuously variable adjustment. A possible way to solve this type of

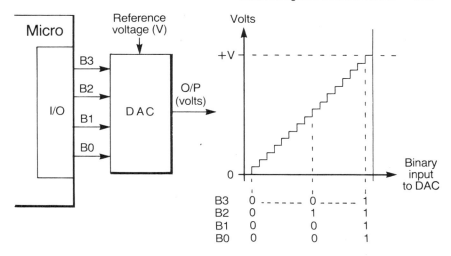

*Fig. 4.9* DAC interfacing and use.

problem is to use a digital to analogue converter (DAC) which is broadly speaking the opposite to an ADC.

DACs are usually found on single integrated circuits, as with most common functions. The internal circuitry of a DAC can be very simple indeed – largely resistors and a single amplifier. However, modern devices are rather more complex than this to achieve greater efficiency and accuracy. Figure 4.9 shows the interfacing and effect of a DAC attached to some of the output lines of a controller.

The output of the DAC gives a voltage which is proportional to the value of the binary number appearing on its inputs – B0 to B3. In this case, the DAC only has four inputs. Practical circuits may have any number, with eight and twelve being typical. The maximum output level is decided by the reference voltage (V).

The diagram shows the effect of inputing the complete range of binary numbers from 0000 to 1111. As can be seen, the voltage range of the output is divided into sixteen equal parts, and as consecutive binary numbers are input, a step occurs on the output. The depth of each step is dependent upon the proportional difference between consecutive binary inputs. Thus, if the voltage V is divided into many more parts, each step value will be much smaller.

To produce a sine wave output from 0 to 360°, for instance, binary numbers for the sines of sixteen equally separated angles from 0 to 360° will have to be fed to the DAC. The output will then step through sixteen levels of voltage over the complete cycle, with quite large steps between each input binary number. The final waveform will be very little like a sine wave as a result. If, however, an 8-bit DAC is used, this gives 256 voltage levels from top to bottom and can be used to give a fairly good sine wave output.

If this full sine wave cycle is repeated continuously from the controller, a continuous sine wave tone will be output, but with the steps still superimposed on the waveform. These steps thus give a high frequency harmonic mixed in with the fundamental. This can then be filtered out electronically to produce a fairly good result.

By using DACs with very high resolutions – say sixteen bits – excellent waveforms can be synthesised from the basic voltage values, in binary form, which make them up. This includes speech and music synthesis. Also, much of our everyday communications are turned into binary form using an ADC, transmitted in this binary form, and then reconstituted at the other end using DACs.

## AN APPLICATION

We will now look at an application based on a project to automate the grinding down of spectacle lenses to make them fit any given lens frame. The application described here is taken from a real project, but the system descriptions, both mechanical and electronic, have been changed somewhat. This is partly for proprietary reasons, and partly to concentrate on practising ideas which have been introduced so far in this book. The last main case-study was of a process control system – data collection – this one is an example of product control. However, many of the development tasks are similar, and will not be repeated below.

The object is to take a circular lens and grind the edge down to fit into any given lens frame. The first job is to take in the shape of the lens frame involved. This is not shown here, but is performed by moving a radial stylus around the groove in the spectacle frame automatically. As the stylus moves around the full 360°, a linear position encoder records the radial position of the stylus, and a shaft encoder reads its angular position. Thus a complete table of radii and associated angles digitises the two-dimensional shape of the final lens. This is then mirror imaged for the other lens in the frame. This data is held in memory and then fed, when required, to the lens edge grinder.

The controller takes this data and controls the machine shown in Figure 4.10. Here, a drive motor with home and shaft encoder rotates the lens to be ground against a grinding wheel as shown. As the lens spins, the whole shaft, including the drive motor, is moved up and down vertically by a linear actuator formed from a stepper motor with a threaded armature, as mentioned above.

The controller starts by lifting the shaft well above the grinding wheel, and the lens is attached to the shaft. The shaft is then turned to the home position, the grinding wheel started and allowed to run up to speed. The stepper then lowers the lens onto the grinding wheel slowly, allowing it to cut until the radius of the lens at that point is equal to the radius measured

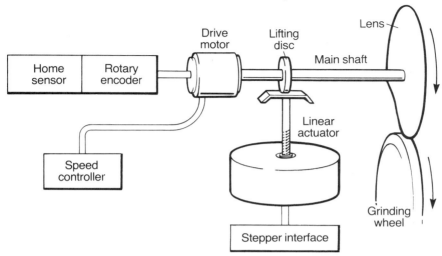

*Fig. 4.10* Spectacle lens edge grinder.

for the same angular position within the groove of the frame. The controller then uses the drive motor and stepper to turn and raise and lower the lens to ensure that the cut is performed to the correct radius as stored in memory. This continues until the lens edge is fully cut. A second rotation then completes the process. In a real machine, a further cut is made on another grinding wheel to leave the edge in the shape of a V to fit into the frame groove.

The biggest problem is that as the lens edge is lowered onto the grinding wheel, it cuts a region out of the lens edge, and over that region the radius will be a convex circle of the shape of the circumference of the grinding wheel. This means that a large degree of calculation and foresight must be used by the controller – particularly for highly curved regions of the lens edge. Consider, for instance, a rectangular spectacle frame and the problem of producing the corners.

The rotary encoder on the drive motor gives continuous feedback of the lens's angular position, and this is tied up with the correct radius for the final lens at that angular position. This radius is used to decide on the control of the stepper. Where the curvature of the lens edge is great, the drive motor must move slowly or the stepper will not be able to change the radius quickly enough, and in any case it is better to perform the cut slowly to ensure that it is performed accurately. This is a fairly complex real-time closed-loop feedback problem.

The development of the control system proceeds as for the data collection example described before. The first task is to produce a specification, as far as possible given that some experimentation is needed before the specification can be completed. The next decision to make is the number and character of the I/O lines, and then the controller must be chosen. In this case, a single board controller of the ARC40 type was applicable, and the number of control lines not great.

The main problem here, as is often the case in a product control situation, was to ensure that the rather complex control principles would work. This required the manufacturer of the machines to produce a lens edging machine for experimentation. They had already performed the calculations as to the size of stepper, etc., using standard motor torque data. By the time the machine was ready, a development system was available to connect directly to the mechanics, and software could then be written to control the mechanics directly.

In this case, the machine manufacturer had no internal computer expertise, and he used an outside contractor to produce the controller. This is a common situation, and proceeds most efficiently when the machine manufacturer takes the responsibility for the mechanics, and ensures that there is no hold-up in its production. It is preferable, in fact, for the machine manufacturer to have this finished and ready before calling in the control designer. However, it may be difficult in many cases, and is a flexible point. If the control designer can be given a finished piece of mechanics to control, he will be able to concentrate on that control, and not be side-tracked.

The experimentation stage of this project was the longest part. Once the problem of control had been cracked, the resulting program was simply stored onto EPROM, plugged into a standard controller, and the system was ready for final production engineering. It cannot be overemphasised how useful it is for the machine manufacturer to have some idea of the meaning of control, and thus make the mechanics as amenable as possible to that control. An understanding of the principles of control also allows the machine manufacturer to understand the full extent to which control of his machine is possible.

## CONCLUSION

This chapter and the last have introduced a few common parameters which are monitored and controlled in machines and processes. The underlying principles are invariant across the field of control of any kind. It is important to remember that a specification can add greatly to the cost-effectiveness of the final outcome, and such a specification should be considered seriously for all projects. It is important to be a knowledgeable user of high technology, and not leave all the important decisions to an outside contractor. The contractor will not understand the process or product as well as the user, and may miss some valuable tricks purely through that understandable ignorance.

It is important to choose the correct level of hardware and software for a given project, in order that the outcome is fully cost-effective. This is particularly true of a medium or high volume controller for a cost sensitive product. It may actually be better to invest in the expensive development cycle of a dedicated unit, with its amortisation over many units, rather than take the initially cheaper route. However, this comes entirely from an accurate assessment of the market for the device, unless that is assured in some manner.

# Chapter Five
# The Operator Interface

## INTRODUCTION

Many projects have foundered in the past because the needs of the operator have not been fully considered. This runs across the gamut of computer applications from large mainframes to small instruments for a given purpose. It is essential to follow certain rules when presenting data to a human operator, and this chapter explains some of them. In addition, the electronics and function of the interface to the operator in a number of different settings is described. This includes simple devices such as we have seen in the data collection example above, and more complex interfaces involving special needs in certain industrial settings.

We will also look at the more advanced communication medium of human speech, and see where it might become important in the future.

As a final practical concept for the design of controllers, the chapter ends with a description of the electronics of multiplexing. This can be used to expand the number of both input and output lines. In a situation where a small controller is to be used with a large amount of operator interface signals, in addition to those of the control parameters themselves, multiplexing will be essential.

## TYPES OF OPERATOR INTERFACE

Everything from a single light on a front panel to a complete VDU and keyboard can be used as an electronic operator interface. Examples of such devices can be seen in the watches on our wrists, announcement boards above station platforms, or the screens and keyboards of personal computers. We will now class the devices to be described here, along with their industrial applications. Some of the electronic principles involved in the devices and particularly their interface to controllers will then be discussed. The emphasis, as usual, is on supplying information to allow the correct level of device to be chosen for a given application.

Table 5.1 gives a summary of the types of interface which will be considered.

*Table 5.1* Classes of operator interface

| Type of Interface | Interface requirements | General interface applications |
| --- | --- | --- |
| Single lights and buttons | Single I/O lines and possibly voltage level change | State indicators and pre-determined commands |
| Audible warning devices (AWD) | Single output lines and possibly power amplifier | Warnings, error indicators |
| Digital displays including LED. LCD incandescent | Several I/O lines plus software and special interface | Aid to operator input, detailed state indication |
| Simple matrix keyboards | Key encoder and several I/O lines | Input of numbers, general commands, setup data and system running requirements |
| Printers including interface | Standard digital interface | Hard copy such as reports and small tickets |
| VDU screen usually with keyboard | Either an internal interface board or a single serial communications line | Can be used on the front panel of process machinery but more often in development |
| Bar code wands | Special interface and software | Automatic data collection for stock control for instance |
| Badge and card readers | Special interface and software | Security and data collection |
| Speech synthesis | Special electronic interface | Process and general control applications |

These are not the only interface examples between machine and operator, but they span a wide enough range to help in deciding on the level required for a given application.

A part of any specification will be a set of requirements for the display of information, and input by the users of the system. For instance, in our data collection system example in Chapter 2, the layout of the front panel, along with its operator keys and display, was described.

Similarly, when a customised production control system is designed, the managers in the company who will use the system will normally be asked to provide a set of requests for output given the type of data which is being handled. The way in which the system is to output information on the production process, and how that process is to be planned is an essential part of the task of analysing the system requirements. It may be that the managers will ask too much of the system, and the systems analyst will have to explain this. On the other hand, it may also be pointed out to the managers that more data than they have asked for can be presented, and it is up to them as to whether any extra cost is worth the increased facilities.

The main point to remember is that the screen layout is often the full extent of that person's experience of the machine. His impressions, and his satisfaction with the system will often rest on this parameter. There is no point in producing the most complex and sophisticated system if the users do not have access to the facilities, and cannot use them fully. This is also true of much simpler systems. It is essential to consider the type of user involved, and how much data he needs to be given.

As an example, a data collection system, with the type of node described in Chapter 2, Figure 2.10, will have to interface with shop floor personnel. They will generally have no interest in the system *per se*, and it is important to make it as robust and easy to use as possible. No extra flexibility need be added, and indeed the clearer and more basic the interface, the better. On the other hand, the production planners may require considerable flexibility and leeway in the system's use, and will be prepared to become fully involved in making the best use of the system. In a large company spending heavily on such computer equipment, there will usually be some member of staff, perhaps with a computing background, who will even be capable of making small changes in unprotected parts of the software, to save continual reference to the system designers.

Industrial systems normally arise from a compromise between the system designers and the system users. It is useful to the designers to have a knowledgeable contact within the user's company. This helps to make the introduction of new hardware and software more efficient. It is often the case that a user with no detailed computer knowledge will either expect too much from a given system, or miss some of the facilities which can be gained. In such cases, a high degree of responsibility rests with the systems analyst and designer.

## SOME SIMPLE INTERFACES

We have already seen some simple operator interfaces, and discussed their use in a system. The most complex of these was the data collection node of Chapter 2. We will look at the degree of I/O complexity required to run this type of interface shortly.

The first two classes in Table 5.1 have already been covered, with applications, to some extent. Simple digital I/O lines will run lights, buttons and on/off devices such as AWDs. It should be said that if a given piece of equipment can use just this level of interface, then it will be easy and efficient to train personnel to use it. The problems with the man/machine barrier are most often met with the more complex types of interface.

Lights can be used to indicate that a machine is on, the exact state of the machine, and whether the operator is required to perform some special action. For instance, when a given machine requires loading with the next piece of material, or requires unloading, a light and/or AWD can be used by the controller to signal this state. Similarly, if the operator is required to load the machine and then start the controlled manufacturing process, a button marked 'machine loaded' can be pressed to signal this state to the controller.

More complex information can be input by the operator by presenting him with a series of switches, each clearly marked with function. This is only slightly removed from the first cases above, and is still not too complex to be picked up quickly and operated with little error. However, it is always useful to provide some automatic background checks on the key pressing, as human error is a continual problem. For instance, if a machine is controlled through a cycle in which the operator has to load and unload, and perhaps move an object through parts of the process, it is useful for the controller to disallow keypressings which it knows are out of sequence. Of course, this philosophy cannot be taken to too great a degree of control without making it simpler to control the whole machine automatically. Nevertheless, it is worth keeping an eye out for obvious opportunities for checking the human operator with small additions to the basic system.

Often, automatic machinery of a simple nature is installed purely to monitor and prevent the human operator from making mistakes. For instance, a gasket maker found itself having to pay out large claims for the serious failure of high compression diesel engines in vehicles using its head gaskets.

Every so often a new engine in one of the larger heavy goods vehicles would fail after ten or twenty thousand miles. On investigation, it was found that one of the gas rings was sometimes missed out, or an extra one included, around the large holes in the gasket. This caused an abnormal thickness change in that area of the gasket, and the resulting strain in an already highly stressed cylinder head would cause the failure.

It was decided that one of two paths could be taken. Firstly, the human operator could be replaced, and the complete process of assembly and pressing of the gasket performed automatically. Careful 100% checks could then be included within the machinery for any abnormalities in the product. Secondly, the existing expensive pressing machinery could be converted to allow an automatic 100% check on the presence of all the parts of the gasket during normal human operation of the process. This would necessitate the machine's prompting the operator to perform each step in the process, and then checking positively that it had been performed perfectly. This path was eventually chosen as being the more cost effective. The resulting machinery presented a few lights and buttons for the operator to use, and was able to save a large amount of money both in terms of the absence of expensive failure claims, and in terms of the comparative alternative cost of complete automation.

This is a good example of the trade-off between complete replacement of expensive machinery, and the modification, for a given purpose, using microelectronics. The problem was solved using a standard computer board from a multi-board computer system. With the board plugged into the computer system the program could be developed using the actual hardware along with its I/O which would be used in the final system. This allowed all the sophisticated software of a large development system to be brought to bear on the development. The final program was stored on EPROM and plugged into the computer board, which, with its own internal I/O, could be used to run the automatic monitoring equipment.

## DIGITAL DISPLAYS

There are many types of digital display, and most of them are familiar in everyday domestic manufactured items. For instance, LCDs (liquid crystal displays) are used in wrist watches, and low power calculators. The power required to run them is low enough that it can be supplied from the incident light on a small solar panel, along with the electronics of the calculator itself.

LEDs (light emitting diodes) have been used for wrist watches in the past, but they consume too much power for this type of application. They are to be found in calculators, process control front panels, car dash-boards, and so on. Their predominant colour is red, though other colours also exist.

Incandescent, or fluorescent, displays use a type of gas discharge not dissimilar to a neon sign or fluorescent light bulb. They are constructed from a largely evacuated tube, and some internal electrodes, and require a high voltage supply. Their predominant colour is green, and where power is no problem they are by far the most easily read. As such they are recommended for process control use where an industrial environment will be found with many different types and levels of ambient illumination.

All of the above displays interface to digital circuitry in a similar manner. There are many different types of display available, and many which will be seen in mass produced objects have been customised to that use. The situation is very similar to that of single chip computers. It is initially expensive to order a customised display, but produces a very low unit cost in high volume.

We will now look at a typical interface for an LED display, and see how the display of numbers and letters may be achieved. An LED is a diode which emits light when a voltage is applied in the correct direction to cause current flow. As explained in Chapter 4, a diode will only conduct in one direction. When current is flowing in the diode it is said to be forward biased. LEDs are rather delicate diodes, and it is easy to cause too much current to flow during forward bias, which results in damage. They normally require a limiting resistor in series to prevent this situation. They can also be damaged easily by too great a reverse voltage across them, and in general it is important to design the system so that they are either correctly forward biased, or off.

When arranged in a matrix, as with any type of indicator, LEDs can be used to produce numbers, letters, or any other pattern as desired. The most common LED displays are known as seven-segment displays, although they usually have eight segments! The seven segments referred to are ones which are capable of displaying the ten digits from 0 to 9. Such a display, with the eighth segment included, is shown in Figure 5.1. This is not necessarily an LED display; it applies to all seven-segment displays. As

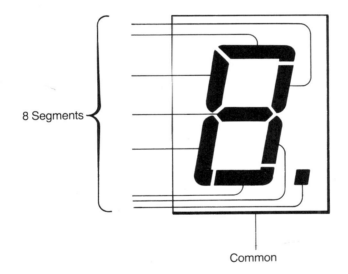

*Fig. 5.1* Digital display layout.

can be seen, the eighth segment is the decimal point.

The connection to the device is by a single line to each segment, and one common line. When a current is passed into one of the lines and out through the common line, that segment glows. All the segments can be on at once, and the common line thus carries the full current loading for the display. By switching on one or more segments, different patterns are produced. Some of these are shown at the bottom of the figure. In general, although single segments are shown in the figure, more than one LED is often used for each segment to increase the light output. Common display sizes are from one tenth of an inch high to one inch high. The smaller displays often include a front lens to increase the readability of the digits.

The display method was first designed for figures and as can be seen, this application is well suited to the seven-segment approach. However, the first six letters of the alphabet are also displayable, as shown, in a mixture of upper and lower case forms. Others can also be formed, such as L and H. This can be useful in some minimal applications, and Appendix 1 shows how important the first six letters of the alphabet alone can be. This provides a very low cost display method, and unless it is essential to have more complex data available, is highly recommended. Simple error messages are also possible on such displays, for instance 'Error' comes out quite well on five digits placed next to each other.

There are also other more complex patterns of digital display using basic lit elements. For instance, the 'star' display includes some diagonal elements which allows letters such as 'K' to be formed. In fact, using a star display, all the letters, numbers and special characters such as asterisk, comma, colon and so on, can be produced, although some experience may be needed to read them immediately.

There are three main ways to interface seven-segment displays to a digital system. Firstly, every single element of each display is connected to its own output line, via an amplifying stage. Digital controller lines do not generally produce enough current for an LED segment without some interfacing. When that line is active, the LED segment attached lights up, and the controller has full control over the displays. This is rather wasteful, and can be helped considerably by multiplexing. This is another example of this important concept, and is explained by Figure 5.2. Many calculator displays are multiplexed, even incandescent displays, and a sure sign of this interface method is to put up a number on the display and wave it in front of your eyes quickly, with your eyes defocussed. The display will be seen to flicker, and the result will be several images produced as the display flashes across your vision.

Figure 5.2 shows how eight outputs from the micro are amplified by display drivers (DD) to connect to the segments – probably one or two chips will be needed. Notice that the drivers may also require external resistors in series to protect the LEDs. These are shown in a resistor pack

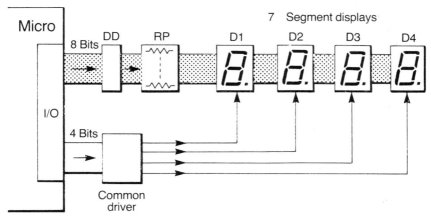

*Fig. 5.2* Display multiplexing.

(RP). However, dedicated drivers normally include these components – they are shown here to emphasise the need.

The diagram also implies that all segments in a given position within each display are connected together, and connected to a single DD line through RP. Thus, all the decimal point lines are electrically connected together, and connected to a single DD line. However, the display commons, D1 to D4, are connected to unique output lines of the micro via four other driver lines.

When the micro wishes to switch on the decimal point, for instance, of a given display digit, it must output two signals. One is the output to all the decimal point segments, via a DD line, and the other is a common output to the given display to be switched on, via the common driver. Thus, even though all the decimal points are activated at once, only one has its common line active at that time, and only that display lights its decimal point. The same is true for all other segments in the digital displays.

To show a pattern on the display such as 1 2 3 4, it is necessary to sequence through all the digits, outputting the appropriate segment drive pattern as each digit is reached. This occurs as follows. First the number '1' is sent to all the displays. None of them shows the '1' yet, however, as none of the commons is active. Then the first display's common, D1, is activated, and the '1' appears there momentarily. D1 is then switched off, and '2' is sent out, followed by a momentary activation of D2. This is repeated across the displays, until it comes around to '1' and D1 again. If this is repeated fast enough, the human eyes cannot follow these high speed changes, and it appears as if the numbers 1 2 3 4 are showing in their correct places on the display – the flickering is lost to the human eye.

The multiplexing occurs on the 8-bit data bus shown. At one moment it carries the data for D1, then D2, and so on. Thus these electrical lines are being redefined as to their use from one fraction of a second to the next.

The economy of this system is seen by considering that there are thirty-two segments here, and just twelve output lines run them fully. This can actually be reduced by placing a 2- to 4-bit decoder feeding into the common driver from the micro, as in the 4- to 16-line decoder which was used in the memory matrix of Figure 2.3. There, four lines were used to signal sixteen lines. As the number of digits increases, it becomes more and more worth while to use further decoding.

An important problem with this type of system is that the micro has to be used continuously for the display, and this can slow the system down. There are two approaches to this problem. Calculators solve the problem by ceasing to attend to the display when they are calculating, and the display often blanks as a result. Another solution is to use multiplexing chips for the displays. The micro simply informs the chips of what it wishes displayed, and the chips perform the multiplexing automatically, freeing the micro for its other tasks. However, in a calculator the micro has very little to do as humans are comparatively slow at using its facilities. There is no problem in simply forgetting the display during its calculation processes.

Another approach, in a situation where a micro must be used as in Figure 5.2, is to service the display on an interrupt basis. A part of the multiplexing routine is performed each time a clock pulse causes an interrupt. The display is thus treated very much as the real time clock of Chapter 2. It should not be forgotten that if the micro is to perform the multiplexing, software routines must be written for it, and look-up tables produced to convert numbers and letters into appropriate patterns of segments to be activated. This can take up valuable memory space and processing time.

Another method of interface, and one which does not require much extra programming, and almost no extra time, is illustrated in Figure 5.3.

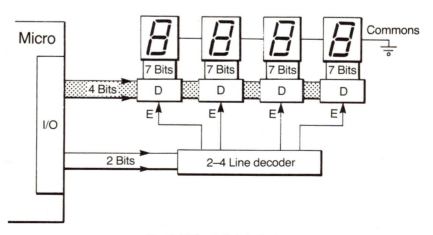

*Fig. 5.3* Direct digital display.

This system has the advantage that the segments stay on continuously, and as such can produce a brighter display. Once set, the controller can forget about them, and they will remain until changed. Also, only six bits are required for the full set of four displays. This circuit does not include the decimal point, but it can be added in easily with a couple of extra output lines and a decoder.

The intelligence of this circuit lies largely with the chips labelled 'D'. These are display driver chips which only require a 4-bit code to drive the segments automatically in the correct pattern for the required display. They contain look-up tables which tie up the input 4-bit code with the correct segments to be lit. Also, all the commons of the displays are wired together, and conventionally connected to the 0 V line. This requires displays with the correct common polarisation.

The circuit works as follows. The micro first selects a given D chip by sending out a 2-bit code to the 2- to 4-line decoder. This 2-bit code is converted into an enabling signal to one of the D chips. Then the micro sends out the 4-bit display code to tell that D chip what it wants to be displayed on its associated display. Each 4-bit line of each D chip is connected through to all the other chips, and thus this 4-bit data bus is multiplexed. Only the activated D chip takes in the data from the bus, thus only one display is changed. When the micro removes the active signal, and moves on to another chip later, the D chip latches (remembers) its output to the display, and thus the display remains constant until actively changed next time. The display matrix thus acts as a visible memory area. In this regime, however, it is not possible for the micro to read back the data on the displays, and it must retain its own internal image of the displayed information if it needs it.

As can be seen, each D chip is presented with sixteen different patterns to display. It is not possible for this system to display any others, as is possible with the last system, and the patterns possible are decided by the particular D chip chosen. It is possible to have special D chip equivalents produced to a given specification, but, in general, standard parts are used. They normally include the ten digits made up as shown in Figure 5.1, but there are still six more possible patterns. Some chips just produce six general graphics patterns, while others provide six useful display patterns for the remaining codes. A useful chip manufactured by Motorola, called the MC14495, includes the letters as shown in Figure 5.1. This allows some simple messages to be displayed, but more importantly for some computer applications, the full hexadecimal number system (see Appendix 1).

The complexity of interconnection of Figure 5.3 is a little more than that of Figure 5.2, but in an industrial application this does not generally matter as a PCB must be made up for the interfacing anyway. The increased efficiency of this system recommends it above that of the previous systems in general.

As we have seen a couple of times, the concept of multiplexing to increase the availability of I/O lines is easy to use, and efficient. We will look shortly at keyboards, and use a similar concept again for the same reason.

To drive other types of display, it is merely necessary to change the power supply requirements, and driver chips. For instance, incandescent displays require a higher voltage than the +5 V which is sufficient for the LEDs. LCDs require a special AC signal to 'power' them, but will work at +5 V, with almost no current.

LCDs are interesting devices, and we will look at these now in case the reader wishes to know how they work, and to show the design considerations which go into their choice in a given setting.

There are several different types of LCD, and their low power requirements make them ideal for battery applications. They have no DC electrical path through them, and thus consume little power. In fundamental terms they are near the ideal method of data communication. They almost transmit pure information without transmitting energy in the same way as LEDs or other displays. Of course, this pure ideal is not fully attainable, and as such some power is consumed, particularly when they are driven in such a manner as to protect them from self destruction. We will see how this works now.

Figure 5.4 sketches an LCD element. The top diagram shows an edge-on diagram of an LCD element. Between two planes, a material is held which reacts to an electronic field. There are several types of LCD, and new ones are being developed continuously. The liquid crystal either reacts to an electric field by becoming opaque, or by changing its light

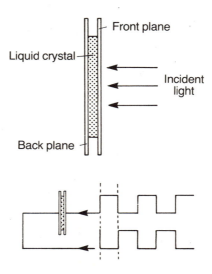

*Fig. 5.4* LCD.

polarisation properties. In either case, as the crystal is activated, it can be seen by reflected light. In the latter case, a polariser is required in the front plane to expose the polarisation change which occurs. Either way, the front plane is clear so that the LCD changes can be seen by its effect on incident light.

In order to apply the electric field, however, both the clear front plane and the back plane must be able to conduct electricity, albeit with minute currents. This problem is solved by vaporising a very thin film of gold onto the surfaces, and connecting these films to pins on the two surfaces. The gold can be thin enough to allow almost complete passage of light through it, but conductive enough to transmit the electric field to the liquid crystal. This is possible as almost no current flows through the liquid crystal itself. If the field is maintained, the element thus exposed remains actively indicating. However, a constant polarisation of field tends to destroy the LCD element, and a continuously reversing field is thus used in general. The LCD acts like a capacitor, and an AC field will conduct some AC current across this capacitor.

The diagram at the bottom of Figure 5.4 shows how the LCD should be driven. A continuous square wave is fed to one side of the LCD, say the back plane. The other side has a square wave either in phase with the back plane, or at 180° out of phase as shown. In this latter condition the LCD is active as it has a field maintained across it most of the time, except when the waveforms are actually changing. A square wave is shown as this is a digital signal which is easy to derive.

Naturally, such elements can be formed into almost any pattern required, and seven-segment displays are quite common. They look like thin pieces of glass with no apparent electrical connection, unless pins are attached to the glass. It is usual either to sit them on flexible conductive rubber contacts, or to clip pin assemblies to them to take up the connection.

There are two important considerations for the use of LCDs in an industrial environment. Firstly, they work entirely by reflected light, unless actually backlit in some manner. This tends to make them hard to see in some cases. Secondly, they usually have a restricted viewing angle, and one often sees users peering into the display and moving around to read the data. This is particularly irritating in the recent development of the LCD output computer. A large LCD panel is used instead of a conventional VDU screen, and all the information normally seen on screen is presented on the LCD panel. The view is difficult to achieve efficiently, and in its present form is not as usable as a VDU or an incandescent display.

It is important to make two further comments in mitigation of the disadvantages of LCDs. Firstly, it is the displays in a computer which normally prevent it from being used with batteries, and the LCD completely solves this crucial problem. Secondly, LCDs in complex applications are usually multiplexed, and this tends to make them 'dim' and thus hard to read. This will be overcome by LCD storage displays where the information can be written to them, and they will be able to retain the data without being refreshed in the same way. An example of such a development is the Standard Telephone Laboratories' 'Smectic A' type of display. However, a suitable commercial version of such devices has yet to be produced.

LCD technology is important, therefore, for battery applications, and for the future of low power computer electronics.

In a process control environment, where power is the least worry, an incandescent or special high brightness LED display is probably the best alternative. However, the LCD is essential for specialised hand held instruments which must run for many hours on batteries which cannot be too large and heavy.

## KEYBOARDS

Keyboards are simply a matrix of switches of one form or another which can be used to transmit data in intelligent human form into a computer. The design and interface of keyboards is a long established art in computing, and in the associated technology of typewriting. The full typewriter keyboard which is common on personal computers stems, initially, from the familiar mechanical keyboard of the early typewriters. Initially, computer keyboards were also rather mechanical. They were usually found along with other electromechanical machines such as teletypes. There, coded signals from the computer would cause solenoids to hammer pins through paper tape, and type-heads onto ribbons to make a mark on the paper. Some of this technology is still used in modern machinery, but with considerable refinement.

We are interested here in common keyboard technologies in process and product control, and this clearly points to the most common type of keyboard, which is the membrane keyboard. The construction of the membrane keyboard has a number of advantages. Firstly, it is easy and cheap to produce even quite large keyboards by this technology. Secondly, the keyboard is very thin and does not suffer from the usual problems of accidental damage by impact of the usual switch-based equivalent. Secondly, it is easy to protect environmentally. The construction is shown in Figure 5.5.

The top diagram shows a membrane keyboard edge-on. The individual switches are just conductive patterns as shown. They are held apart by the spacers and the natural elasticity of the plastic membranes on which they are printed in conductive ink. Finger pressure brings the conductive pads in contact, and they spring apart when released. Part of the pattern of switch pads on one of the planes is shown at the bottom. The pads are

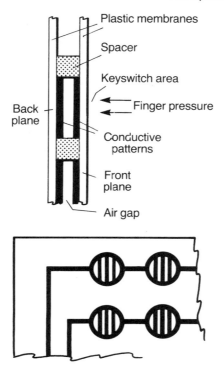

*Fig. 5.5* Membrane keyboard.

connected in rows on one of the membranes, and in columns on the other. As the whole manufacture is one of photography and printing, the front membrane can be printed with any type of graphics required. It is not expensive to have a customised keyboard produced in quantities as low as 50 or 100. Alternatively, a standard keyboard matrix can be purchased, and a thin plastic printed cover stuck to the front for a customised legend.

Despite the apparent simplicity of this technology, which depends upon plastic membranes springing apart, such keyboards rarely give problems, and will last for a remarkably long time. They stand up to a wide range of temperatures and humidities, including, in the author's experience, continuous wiping with a wet cloth and constant steamy conditions in the food processing industry. If the conditions are very hard, they will fail every so often, but replacement is extremely simple and cheap, and can usually be left to the purchaser of a process control system, as part of the normal maintenance procedures.

The connections normally come out along a ribbon of the membrane's material to a standard ribbon cable socket, and can be plugged directly into the electronics. An important parameter to note about the membrane keyboard is that it is less positive than a normal keyswitch-based matrix, and it is up to the electronics and control software to allow for this. Keyboard 'bounce' is particularly marked. This refers to the fact that a single keypressing of a membrane may cause many closures to be noted by the electronics – to this extent the switch is electronically noisy. It is also of a higher resistance than normal switches. We will now look at interfacing the membrane keyboard.

Keyboard interfacing is fairly similar across the board, and we will look at two typical methods to assess how much I/O is required. A keyboard is a matrix of switches, and rather than connect each switch to a separate I/O line, which would be wasteful, the switches are arranged in a form which can be visualised as a matrix of rows and columns, as shown in Figure 5.6. Each switch connects a given row line to a given column line. The figure shows that the controller has the job of decoding which switch has been pressed by looking at these rows and columns. Four inputs and four outputs are required, as shown.

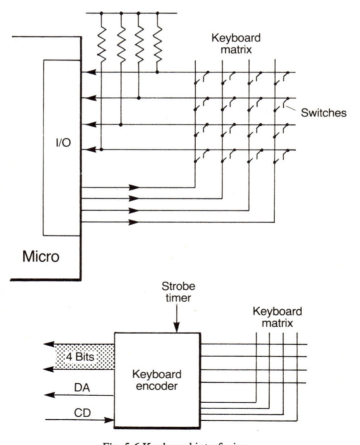

*Fig. 5.6* Keyboard interfacing.

To see how this works, imagine that the output from the controller connected to the left most column is outputing a '0', and the others are all at '1'. The controller reads all the inputs (rows) looking for a '0', but does not find one. To ensure that this is the case, each of the otherwise floating inputs is held up to a '1' level by a resistor. If a switch in the left hand column is now pressed, one of the row inputs transmits the '0' from the column through to a row, and the controller can tell from that row which switch has been pressed – it already knows which column is connected to the switch because that is where the controller is outputting a '0'.

To ensure that all the switches are scanned, and no keypressings are lost, the single '0' which is being output is moved or 'walked' along the columns. Meanwhile, the rows are read for any zeros which may appear thus signalling a keypressing. This method of decoding is called 'walking bit decoding'.

If the scanning frequency is high enough, the normal human speed of pressing keys will not defeat it. In fact, any given keypressing will cause multiple readings of that key just from the decoding method, let alone any shortcomings of the mechanics. It is up to the software to be intelligent about its reading of the keys. It will have to look for a given period between the keypressings of any given key before it takes it as a second pressing of the same key. Sometimes, such second pressings may be missed, and it is up to the operator to take care. This leads to the necessity in almost every case for a digital display, or some form of feedback, to assure the operator that his input was correct. This has already been mentioned in the data collection system example in Chapter 2.

If you examine the matrix of Figure 5.6 with care, you will be able to see that if two keys are held down at the same time, two of the controller's outputs will be connected together. This is not a desirable situation, and it is normal to use diodes and more resistors to defeat this possibility. In addition, it may be necessary to use a further electronic interface between the keyboard and the controller to ensure that the high resistance of the contacts does not affect the reading.

In fact, direct interface to the micro in this form is a little inefficient from the software point of view too, as the micro has to contain the 'debounce' software, plus a separate I/O line for each matrix line. The design at the foot of the figure shows a common solution to the problem, and once again shows how external hardware, in an IC, saves components, processor time and extra program development which may be more expensive than the hardware addition. However, this latter point may not be true in a low cost high volume product. There every component adds greatly to the overall cost of each batch, and a little extra development cost will quickly amortise out to a zero cost-burden on each item. The extra cheap passive components may also be preferable to a more expensive integrated solution. As usual, correct design choice is a compromise.

The lower diagram shows the principle of using an integrated keyboard encoder. It connects directly to the keyboard matrix, and converts each of the sixteen switches shown here into a 4-bit code. The internal electronics of the encoder generates the walking bit data, includes the pull-up resistors, and compensates for high contact resistance. In general, only one key can be detected at any time, and no code is generated until the key has been closed for some time.

This delay time allows for any closure bounce. The time value is decided by the strobe timer, which will normally just require a couple of passive components (resistors and capacitors). The delay time is normally adjustable by the component values. When a key closure is detected, the 'data available' (DA) line becomes active, and this signals that a key code is available for reading into the controller.

To signal that this reading is complete, the micro then sends a signal along the 'clear data' (CD) line. This resets the DA line, but usually leaves the 4-bit code active. However, it is up to the controlling program to ignore this until a new DA signal is given. This DA line could be connected to the interrupt line of the MPU, and thus the micro would simply ignore the keyboard until an interrupt occurred. An interrupt routine would then read in the key, and hold it until needed by the main program.

As can be seen, just six bits are needed for this process, and considerably less software. Again, the micro may hold a look-up table which ties the 4-bit codes up with keyboard key values for comprehension by the main program. A useful advantage of a look-up table, even if not strictly needed by the software, is that changing the meanings of the keyboard is a straightforward process of changing the look-up table values. This may not be relevant in a case where the keyboard is numeric, and always used entirely for that purpose. However, in a process control application where there may be several nodes strung around a shop floor, the more standardisation and ease of adaptation included the better. This is particularly true for the first leg of an ultimately expandable system.

The principles of keyboard interface shown here transfer to any type of key switches. The larger typewriter keyboards are often decoded in this manner too. These larger keyboards, used on computers, are often called 'QWERTY' keyboards to indicate that they are similar to a typewriter, which has these letters at the top left hand side of the keyboard layout. Such keyboards are also called 'ASCII' keyboards. We shall see the meaning of ASCII in Chapter 6, when we look at communications and the associated methods by which the letters and words of the operator's natural language are usually encoded within computer systems.

## The application of keyboards

In control, the term 'keyboard' includes almost any collection of switches which the operator may be able to press on a front panel. We have seen some general examples of such devices, and a detailed example in data collection. It is important to stick to some basic rules in deciding on a keyboard layout in a given situation.

As mentioned before, it is good sense to keep the keyboard as simple and easy to use as possible. This does not just mean few functions – it also bears the environment in mind. In the example of a rugged environment where the users will probably have gloves on, it is important to space the keyswitches well apart, and make them as large as possible. This can be achieved with membranes or individual switches set in a metal case. It is also important to protect the switches themselves from being knocked and from the ingress of fluids and dirt. Again, this can be achieved with careful design by either keyswitch technology, but is easier and cheaper using membranes. The only major objection made in this latter case by some operators is that membranes do not have a positive action – it makes the 'feel' less satisfying. This can be overcome by using a membrane switch which includes 'click' discs within the pads, but in general a little experience normally overcomes this minor objection.

Another essential point to remember is that if numbers, for instance, are being input, a positive feedback of the numbers keyed in is needed. This is why keyboards are normally associated with displays. The controller reads the keypressing and echoes it back to the display. The operator watches the display and when he is satisfied with the input, he presses an 'ENTER' key to indicate this to the controller. In a simple situation where a very few dedicated key functions are presented on the front panel, this may be unnecessary, or a light next to each key can be lit to indicate that the associated key has been pressed. In non-membrane keyswitches it is possible to include a light within the switch top itself for this purpose.

When designing a keyboard for your application, therefore, keep a mental image of the final operator and his environment in mind throughout.

So far we have followed Table 5.1 in examining operator interfaces. The next section there is printers. However, this is best left until standard communications protocols are considered in the next chapter. We will now look at VDUs.

## THE VDU SCREEN

The term 'VDU' stands for visual display unit. It is normally applied to a TV-type cathode ray tube (CRT) screen, and usually associated with a full QWERTY keyboard. This is due to the fact that many types of larger computer system use VDUs with integral keyboards as their I/O peripheral for operator communications. When standard communications protocols are examined in the next chapter, the usual interface method for such peripherals will be described. However, we are interested here in the

VDU screen as an operator interface in process or product control.

The provision of a VDU screen is not usual with small control projects as it requires a fairly large amount of electronics, and will be comparatively costly. In cases where simple numbers and a few letters are required, it is also less confusing to use a digital display as described above. However, there are places where a screen, with its sophisticated communications ability, is necessary. In these cases, the interface is important, and we will look at this now.

There are two methods of interfacing a screen to a controller. One is to treat the screen as a normal VDU and interface it as usual in computer

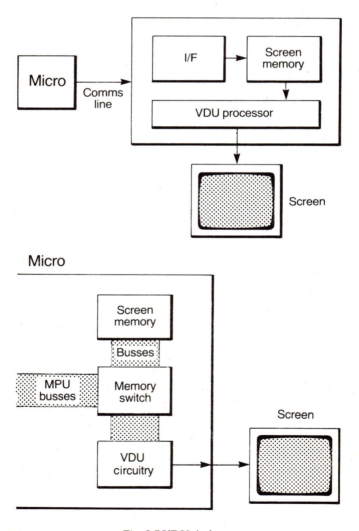

*Fig. 5.7* VDU design.

applications, via an 'RS232' interface, as described in Chapter 6. In this case, the VDU itself will be almost as complex as a complete micro system, as it will have to make sense of simple 8-bit codes sent to it in a standardised manner, and manipulate the screen accordingly. The other method of interface is called memory mapping. Figure 5.7 distinguishes between these two concepts.

In the top diagram, a VDU external to the micro receives data along a standard communications (comms) line through its comms interface. The data is sent to a memory block (screen memory) and the VDU processor simply scans through this memory throwing the data up on the screen in human readable form. As the scanning takes place, the memory is being updated from time to time, and this automatically appears on the screen. There is further electronics within the screen section itself to take the electronic pulses from the VDU processor and convert these into drive voltages for the CRT. No direct connection exists between the internal buses or memory map of the micro, and that of the internal electronics and memory map of the VDU.

In the lower diagram of Figure 5.7, the screen memory is part of the MPU's normal memory map. It is addressable in the same way as any other part of that map such as RAM, ROM or I/O. It is, in fact, a piece of ordinary RAM. However, some VDU circuitry is continuously scanning this memory and, as before, throwing the contents up on the screen via the screen's electronics. The problem comes when the scanning circuitry and the MPU wish to access this block of MPU RAM at the same time. It is solved by using a memory switch which allows the MPU to access the RAM. This causes a slight problem in that if the MPU wishes access while the scanning is in the middle of the screen somewhere, it can cause an unacceptable blank region there. The MPU, therefore, is electronically denied access until the current video line on the screen is complete. This does tend to slow the processor down considerably, and is the reason that programmers prefer to access the screen memory as rarely as possible, consistent with displaying all the information.

The result of the VDU electronics is to allow a window into the internal memory of the MPU system in some region of its memory map. This is where the term 'memory mapped' is derived. This type of VDU design will usually produce a much faster screen response than the other where data has to be sent byte by byte along a comms line before the screen can display it. This is particularly important in such applications as mimic displays where a visual display of a process may be animated in front of the operator. This may well require a fairly fast system.

VDUs are used when complex and sophisticated data would be difficult to cram into a set of digital displays, and a resulting multitude of I/O lines needed. For instance, in the thermoformer case, the two banks of heaters could be shown on the screen, with the heater pattern in real time. A coloured screen would even allow the temperatures to be shown vividly in coloured regions, though different scales of grey on a monochrome screen would be adequate. It is also important to show large numbers of timing periods and interlocking delays on large machines of that kind. This has two aspects – their display and their input by a setter. Again, a VDU and keyboard would condense the problem considerably.

A VDU system on a large controller could show a complete representation of the operator's manual, or the real-time state of any part of the machine, including breakdowns and faults. There is no end to the complexity of data which can be presented. However, this is expensive to produce and should be considered carefully before including it. Also, care must be taken not to demand too many complicated facilities. The hardware is not too expensive, but the software may be prohibitive. The possibilites are also restricted by the type of VDU chosen. If graphics are required, for representing a large process, beware of the programming. The software required for general graphics is involved and expensive.

## OTHER INTERFACES

Table 5.1 also includes bar code wands and badge and card readers. The latter are useful in security or clocking-on systems. Each operative is issued with a badge or card containing a magnetic or other machine-readable code. This is used to identify the individual, and collect his presence at a given position and point in time. The type of card involved is similar to that used in the automatic cash dispensers and bank transaction machines with which we are all familiar. Bar codes, on the other hand, have found most of their use in point of sale systems or automatic stock control. Specialised bar codes are also used in automatic data collection systems, as described in a previous chapter.

We will look at how bar codes work, and their general use in industry. Figure 5.8 reproduces a standard type of bar code found, in this case, on the side of a breakfast cereal packet. As can be seen, there are areas of dark and light, and underneath a human-readable number which the bar code represents. The code is read by a 'wand' in the form of a pen. The tip of the wand has to sense the areas of light and dark as it is moved over the bar code, in either direction, from one side to the other. It does this by shining the light from an LED down through the tip, and detecting the reflected light by a light detector. Naturally, the best conditions are when the tip is directly over the bar code, and when the bar code is in black and white. However, the wand must be usable at a variety of angles, and must be able to read codes of a variety of colours, depending upon the graphics used by the product manufacturer, who may not wish to use black ink on his product packaging.

These are the electronic problems of detecting the difference between dark and light areas. The next problem is the decoding itself. As can be

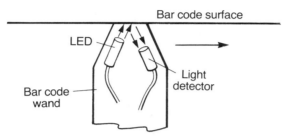

*Fig. 5.8* Bar codes.

seen, bar thicknesses vary, as do the thicknesses of the white space. It is by detecting these different thicknesses that the controller reads the bar code. The general methodology involves first building up a picture of the bar thicknesses in memory.

To do this, the controller samples the output from the wand in a very fast time frame. As the operator moves the wand over the code, the controller is continually reading and storing the light value transmitted by the wand. It takes everything above a certain light value to be white, and everything below to be black. The contrast ratio of the bars affects this process considerably, of course. As the wand moves, the controller can be thought of as pulsing the wand for information. As the reading pulses are very fast, even though the light and dark areas are thin, many light pulses will be picked up from the wand before the dark area is entered. The final memory contents of all these pulses give a proportionate value to the thicknesses of the bars, as long as the operator has wanded the bar code at a constant speed. This is another source of problems and, in most systems, the operator has to learn the 'knack' of inputting data.

In the above system, it is only the ratios between the different thicknesses of lines which are important, and thus the whole code can be magnified or reduced within reason as the manufacturer desires. There are agreed standards of graphic reproduction of the codes, and the system is expected to perform on codes produced to within these limits.

The grocery trade has now conformed almost across the board with the need to bar code its products. This means that supermarkets, for instance, can simply input the bar codes on incoming goods, instead of entering each item on each delivery by hand. An automatic stock control system thus knows what items have been delivered. The loop is closed either by having operators checking the shelves for current stock continuously, or by giving the checkout desks bar code reading equipment so that the stock control system knows what has been sold continuously.

By this means, a computer system can give the buyers instant information on the items which are needed, the items which are selling well, and those which are doing badly. By tying in the accounts too, profit and loss overall, or item by item, can be deduced automatically.

These principles are immediately applicable to a manufacturing industry where raw product of every variety is brought in and the product is bar coded. This means that at least finished stock can be recorded accurately, even if raw material is not coded.

An application which has used bar coding successfully is in the storage and production of spectacle lenses. This requires a very large stock of basic lenses to be kept, and it must be updated continuously. This involves the warehousing of rack upon rack of bins containing lenses. The system adopted was to place bar codes on the outside of each bin. When a set of prescriptions is collected, a portable controller with a bar code wand is used to wand in the code of each bin from which the lenses are collected. In addition, the operator carries a standard set of bar codes for the numbers from 1 to 50, in order to allow the quantities selected from the bins to be recorded.

Thus if three lenses are to be collected from a given bin, the bar code of that bin is wanded in, followed by the bar code for '3'. This is repeated until the collection is complete. At the central station, the controller is plugged into a computer and this data captured for use by the stock control system. A similar process is used to input new lenses into the bins. By this means, the stock control system has an accurate and complete picture of the warehouse to ensure that new lenses can be ordered from the shop floor as required. This type of application would fit a number of different storage settings.

An extension to this system was suggested to incorporate the lens edging machine into the system. In the cases where there are standard lens frames, the internal control number of each standard frame could be stored onto a bar coded label, using a standard bar code printer. This could then be attached to the standard frame. The lens edging controller would have some peripheral memory where it stores data of the cutting pattern for all the standard frames, and it could have a bar code wand attached to it. When a lens is to be cut to this frame, the lens edging controller could wand in the bar code on the frame, recognise the number, call up the cutting data and cut the lenses accordingly.

Bar codes are usable in any situation where there is the possibility of error in the input of data. This is common in most situations where a large amount of similar data is to be input or written down by hand.

The electronics of the bar code interface are not difficult. There are also standard wand tips which integrate an LED and light detector as required for the wand tip. The problems are rather more those of reliable decoding in the controller, as explained above.

## SPEECH SYNTHESIS

This area of modern electronics has been, to some extent, the 'Cinderella' of modern computer technology. Everyone agrees that the synthesis of human speech is an important step towards ideal computers; few have successfully applied it to industrial situations. It has been left aside to be used by a successful toy industry.

The first attempts to synthesise speech, or at least trick others into believing that it had been done, can be found in classical times. More recent attempts in the last few centuries have produced a wealth of clever mechanical devices, all of which require the immensely complex control system of the human being to work them. Even in this century when the first electronic speech synthesiser was presented to the world at the 1939 New York World Fair, it took the operators nearly a year of full time training to learn to operate the controls to produce understandable speech. This first attempt merely used electronics to produce the basic sounds, and not for the more complex problem of the control.

The first successful electronically controlled synthesisers were stored speech systems, as are many of the cheaper systems today. Such systems store words recorded through a microphone, in a compactified form, and call them up by addressing, as with any other form of memory, when required. The complexity often resides in a certain amount of hand crafting of the stored data.

The uses of a stored speech system are immense. Each separate word is instantly and randomly recallable, and the words can be placed next to each other, in any order, to form new sentences. In its ultimate form, a synthesiser would have to store tens of thousands of words, and it would soon run out of memory. However, as memory becomes more and more compact, this will cease to be a problem.

The other approach to speech sythesis is to allow text to be typed in directly. The underlying rules of the words are used to produce their sounds from phonetic components, or 'phonemes', stored in memory. This is quite difficult, and in English, for instance, requires the storage of a large number of special cases if the words are input using their natural spelling. However, a practical use of such devices is to input the sentences using a phonetic spelling. There are several such systems around, and experimentation can usually be performed on small personal computers by purchasing cheap peripheral boards.

The simplest form of speech synthesiser for industrial applications, however, remains the stored speech system. A few standard words are held in a ROM, and called up when needed.

An example of the use of speech synthesis which has been produced for a non-industrial setting is the speaking depth sounder. A problem for boats in some situations is that a continuous readout of depth is required as a tricky manoeuvre is being made which uses the eyes and hands to the full. This can be achieved via the one sense which is still free in this situation – hearing. The sounder simply calls out depth at intervals whose timing can be varied as required.

Such an application is typical of any process where the operator has to use his eyes and hands continuously while watching yet another parameter with great care. The solution might well be to read one of those parameters into the sense of hearing.

The philosophy of stored speech synthesisers is to provide a few of the most useful words for a given situation. The numbers, plus 'please' 'yes' 'no', and so on would be suitable candidates for a synthesiser. Some special words would be needed for the depth sounder such as 'fathoms' or 'metres', 'too shallow!', and so on. In an industrial situation, such words as 'tons', 'amps' and so on may be required. In a security system, the sound of dogs barking may be useful or 'stop thief', which has been used in the past.

The electronic interface of a stored speech unit is normally very simple. Such a system is shown in Figure 5.9, where the usual type of handshaking is used. Here, the speech synthesiser is connected to a bank of ROMs

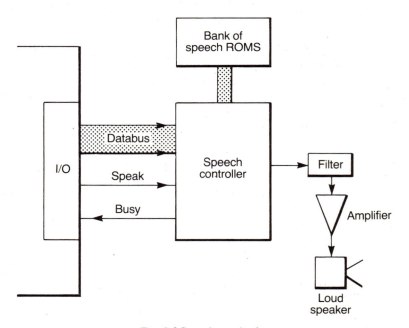

*Fig. 5.9* Speech synthesis.

containing the speech data. Its synthesised speech may be output via a filter to make the sound a little more palatable, and an audio amplifier.

The micro may use a look-up table to connect a binary code with each utterance held in ROM. When it requires one, it places the code on the output data bus shown, and checks the 'BUSY' line from the synthesiser. If this is active, then the last utterance is still in progress, and the micro waits until the speech system is free. Then it commands the synthesiser to speak the word associated with the code on the data bus by signalling along the 'SPEAK' line. BUSY goes active again, and the speech is produced.

In some synthesisers, it is also possible to vary the tone of the stored speech, and thus accent correctly for context. By this means a very natural sound can be produced, but this is not particularly important industrially. It is only essential to be intelligible.

The system shown in Figure 5.9 is not expensive; it uses mass produced chips, including some standard words already in ROM. New words can often be bought if already coded, or taken from a high quality tape if desired. In this way, very personalised speech can be included in a given system.

The figure shows how just two control lines are used plus a data bus. A 6-bit data bus will allow 64 utterances to be coded. Each utterance can be as long as desired, within the upper limit of the ROM memory capacity. Thus if complete phrases are required, they can each be given a code and called up in full without recourse to individual words.

This also provides a quick and simple method for a man/machine interface. It can help to break down the inevitable barrier between operator and machine. It is still immense, and perhaps will only be fully removed by the truly intelligent machine which can converse in normal human terms. This is still some way off, however, and meanwhile as much as possible must be done to break down these barriers by creative design.

## MULTIPLEXING

In order to prevent any restriction on the amount of operator interfacing used in a control project, the number of lines of a small controller will normally need to be expanded. The operator interface is naturally slow, and the time slicing introduced at the end of Chapter 3 will be faster than can be detected by the human operator. This means that it will appear as if all the operator interface signals are being treated simultaneously. We will now look at the type of electronics which is used for this important purpose by considering a system with a few I/O lines, perhaps provided by an LSI I/O chip.

When expanding the I/O lines of an LSI I/O chip, some of its lines are used to feed I/O signals to a variety of outside world lines by redefining the use of the LSI I/O lines from time to time. We have already seen an

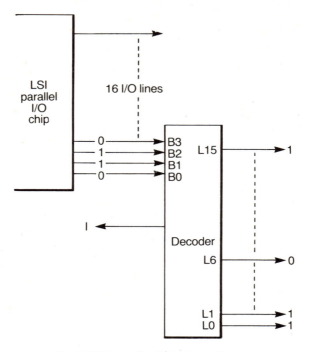

*Fig. 5.10* Expanding I/O using a decoder.

electronic block which can achieve this aim in Figure 2.3. This was the 4-to 16-line address decoder. We will now look at how four output lines from an I/O chip can be expanded into sixteen using such a block. Figure 5.10 shows the electronics involved.

The four outputs chosen are the four least significant bits, B0 to B3, of the chip's I/O lines, all programmed to be outputs. Figure 2.4 and its associated explanation gives an example of the type of I/O chip involved, and either the B-side or A-side may be considered here. These bits are fed into the inputs of the decoder, and all but one of its outputs are at a 1 level. The decoder outputs are labelled L0, L1 etc., up to L15. Disregard the line labelled I until later.

The diagram shows an example of an input into the decoder where B0 and B3 are 0s, and B1 and B2 are 1s. This is the binary number for 6 (in decimal) – see Appendix 1. The decoder is designed to give a 1 on all outputs except that output numerically equal to the input binary number. Here, the input is 6 and thus L6 has a 0 on it. In this way, the MPU can output one of the sixteen possible binary codes on B0 to B3, and thus influence the output pattern of the decoder. This gives the MPU an extra block of sixteen outputs for the use of just four I/O lines of an LSI I/O chip. Using all sixteen lines of the I/O chip in blocks of four in this way, will allow 64 outputs to be produced from four decoder chips.

It should be clear that there are restrictions to the sixteen lines of each decoder's output produced here. Only one decoder output can be at a 0 level at any instant, thus a general 16-bit binary pattern cannot be output. This would require a different type of 'clocked' decoding system, as described next. However, any one of 64 machines or other controls can be switched with this system, and it provides a very low cost method of expanding the I/O structure if the restrictions can be tolerated. A further refinement can be added to the system whereby an extra ouput line from the LSI I/O chip can be used to enable the decoder. In this manner, it can have 1s on all of its outputs except when the enabling line is active, at which time the designated output gives a 0 as above.

Input can be handled in a similar manner using a 16-bit multiplexer chip. This is the reverse of the decoder shown. Imagine lines L0 to L15 being inputs to the decoder instead, and B0 to B3 being selectors. This time, the binary pattern on B0 to B3 decides which L line will pass through the decoder to the line labelled I. This allows the MPU, via its I/O chip, to read any and all of the Lines L0 to L15 along line I, which will be connected to an input of the LSI I/O chip. The line I is thus truly multiplexed – it behaves as any of the L0 to L15 lines depending upon the selection code on B0 to B3 . This code essentially redefines I as required.

As a final note to show the power of multiplexing, we will look at a 'clocked' system to show how complete binary patterns can be output on many parallel lines given just a few. Figure 5.11 sketches the system. Here, four I/O gates are fed from eight lines of the B-side of an LSI I/O chip. These gates are able to pass binary patterns to and from the B-side lines depending upon the state of their E pins (enables) and the R/W line (read/write). The I/O gates are all connected to the eight B-side lines as if

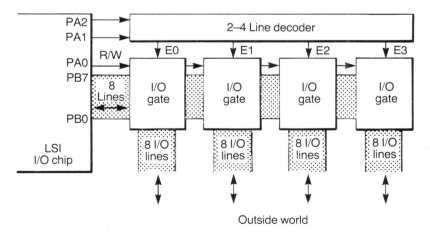

*Fig. 5.11* Clocked I/O expansion.

these lines formed a data bus – this is denoted in the diagram by the B-side lines running 'through' the schematic blocks of the I/O gates. When the R/W line is in one state, the I/O gates are all inputs from their eight I/O lines to the B-side lines. When R/W is in the other state, they pass the B-side lines through to their I/O lines as outputs. Only one I/O gate is active at any time, depending upon which one is activated by its E pin; the E pins are labelled E0 to E3. These in turn are fed from the outputs of a 2- to 4-line decoder which simply turns the input binary state on PA1 and PA2 into an activating signal on a unique E line. We say that each I/O gate provides an 8-bit I/O 'port'.

In this system, thirty-two I/O lines are 'created' from eleven I/O lines of an LSI I/O chip. The MPU sets up a pattern on, or reads the pattern from, the PB0 to PB7 lines, and then 'clocks' or activates the required I/O gate and its I/O port to send or receive the eight data bits.

If the full A-side of the LSI I/O chip were used to feed a 7- to 128-line decoding system, then each of its decoded outputs could activate a unique I/O gate having eight I/O lines as shown, and a maximum of 1024 I/O lines could be serviced directly. In addition, if the I/O gates have latching outputs, then the MPU could simply run through setting each 8-bit output port to any required binary pattern, and these patterns would be held there until actively changed later. This is exactly the same as for the LSI I/O chip outputs themselves.

This system can, in fact, be multiplied up *ad infinitum*. Each pair of I/O ports could be used to provide a further 1024 lines of I/O, and every group of sixteen of those could go further, and so on.

The disadvantages of this method in control are twofold. Firstly, the system provides slower I/O lines than those on the ports of a standard LSI I/O chip. This is because several operations have to be performed in addition to the regular operations of reading and writing the LSI chip. This sequence includes setting the R/W line and the A-side lines to select the appropriate I/O gate before the data transaction can occur. However, this is not usually a problem unless further stages of I/O are included as described for expanding above the 1024 possible. However, this speed problem must be watched in time-critical systems. The second disadvantage is that extra hardware is required, perhaps on an external interface board to take the extra I/O gates. However, this can be brought to a minimum by using LSI I/O chips themselves for the I/O gates. This usually provides the most cost-effective approach to this form of multiplexing.

An example of the use of this type of system might be in a case where a small single board controller has been chosen, perhaps with only sixteen available I/O lines for cheapness, and expansion of the I/O lines is needed. It is not difficult to use this principle to run, say, two extra LSI I/O chips external to the SBC with perhaps twenty I/O lines each thus giving a total of forty full I/O lines for a small sacrifice of speed.

It is even possible to do without the 2- to 4-line decoder in Figure 5.11 if the four lines E0 to E3 are simply taken from the unused A-side lines of the LSI I/O chip. This uses thirteen lines instead of eleven, but dispenses with one of the interface chips and its associated interconnection.

It is perfectly possible to run any type of I/O chips in place of the I/O gates by this method. For instance serial communications can be added to an SBC having no such facilities by using the appropriate LSI I/O chip in place of the I/O gates shown.

# Chapter Six
# Data and Signal Communications

## INTRODUCTION

The communication of data within a system is at the heart of the function and make-up of many controller designs. This ranges from the communication of simple keypressings on nodes in a data collection system to the intercomputer communications within a large network.

The applications which concern us here are communications within process and product control systems. This involves signals over long lines, data sent to printers and other computer peripherals, and standard protocols used in data communications. These protocols are often much misunderstood, and this chapter will briefly introduce the most popular protocols, and describe some of the devices which are used for their implementation.

## CLASSES OF DATA TO BE COMMUNICATED

As usual, the aim is to place the types of communication within a system into context. We have seen some examples of communications within systems already, and others have been left until this chapter. Table 6.1 shows the types of communication which will be examined here.

*Table 6.1* Types of communication

| Data Type | Interface method | Application |
|---|---|---|
| Sensor signals | Level or digital conversion | Feedback from a process or product |
| Operator signals | Level conversion | Process control in a physically extended system |
| Computer or digital data | Serial or parallel | From one MPU system to another or to a printer or other peripheral |

## SIGNAL COMMUNICATION

Examples of the communication method for the transmission of sensor signals were seen in Chapter 2. For instance, the problem of sending low level thermocouple outputs from one place to another was solved by converting voltage to the frequency of a set of pulses. Pulses can be sent over a long line with less chance of degradation than small analogue signals.

The problem of transmitting the closure of a microswitch from somewhere in the middle of a large machine, to the controller, is solved by using a high switched voltage. This tends to swamp the ambient electrical noise. A simple voltage converter, and perhaps an opto-isolator at the controller end will interface this high voltage to the low internal logic voltages.

In general, it is easier to transmit digital signals over a distance than the equivalent analogue signals. This is simply because a digital signal is either on or off, or present or absent, and should have no grey area to be lost in the noise. Care is still needed, of course, due to the possibility of AC pickup, but this can be eliminated by fairly standard electronic design.

Having said this, there are some practical limits to these basic communications methods. If there are very long distances involved, it may

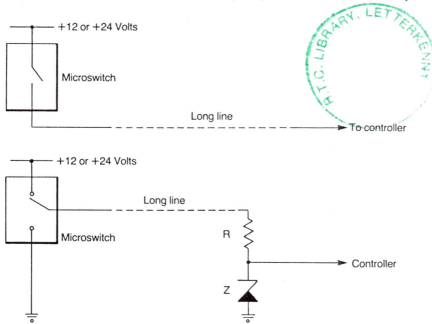

*Fig. 6.1* Switch signals over a long line.

be preferable to use fibre optics to transmit the signal. If there are many variables to be sent, it will be expensive to provide many separate lines, and it may be best to treat the data a little more intelligently and use a full serial communications system over a single line. We will look at serial systems shortly.

Some methods of transmitting a microswitch signal over a long line are illustrated in Figure 6.1. The two diagrams show a microswitch, perhaps somewhere on a shop floor machine, whose state is to be monitored by a controller. It is crucial that the switch sends a definite signal along the long line when in either state. A relatively high DC voltage is fed to the switch, 12 or 24 V for instance, in order to swamp any noise and produce a definite drive to the receiver at the controller end.

In the top diagram, when the switch is closed there is no problem. The DC voltage is fed to the controller and will give a definite signal at that end. If, however, the switch is in the open state, the long line acts as an antenna for all the AC noise in the environment, and anything is possible at the controller's end of the cable.

It is crucial that all such switches have two definite states – any floating state will allow the ingress of noise. The second diagram illustrates the correct approach in this case. Each switch is fed the DC voltage on one side, and connected to a ground on the other. It is important, of course, that this should be physically the same ground as that at the other end of the line. At the controller end, an opto-isolator can be used to ensure that the controller is not subjected to undue electrical strain from the controlled machinery. A simple receiving circuit is shown which will convert the high DC voltage to logic voltages or to the correct voltage for an opto-isolator if it cannot handle the DC voltage level used at the microswitch.

The incoming voltage is passed through a resistor, R, and appears at the top of a 'zener' diode Z. Z has the property that when a voltage appears at the top as shown here, it draws more and more current from the voltage source until the voltage is forced to drop to a given value depending upon the physical make-up of the zener. There are zener voltages for all values from a couple of volts upwards. If a 5 V zener is chosen, when the switch is closed a 5 V signal appears at the controller line shown.

The resistor R is essential to ensure that the zener does not try to draw too much current from the DC power supply, and overheat in its attempt to reduce the applied voltage. R must not be too high a value of resistor, or too little current will be drawn, and any noise on the line will be sufficient to make the zener act to produce a signal to the controller. To state it in a technical manner: the system must have a low 'impedance'. This means, roughly, a low resistance to current flow in order, broadly speaking, to leak away noise signals easily, but still transmit the high current signals from the microswitch.

When the microswitch is in one of its positions, the long line is grounded through R. This is not entirely satisfactory as noise spikes only have to be of the order of three or four volts to produce a signal at the controller end. In many ways it is better to use a double power supply at the microswitch end and switch +24 V in one direction, and −24 V in the other. The receiving circuit has to be a little more complex, but this can overcome problems encountered with the above system. However, the system shown in Figure 6.1 is a cheap and fairly safe system to adopt in many cases, and can be considered in low cost systems, at least experimentally, until proven unsatisfactory.

More sophisticated methods of transmitting signals will use special balanced lines where an attempt is made to overcome noise by cancelling it out. Also, more sophisticated receiver systems can be constructed using op-amps and special sender circuits.

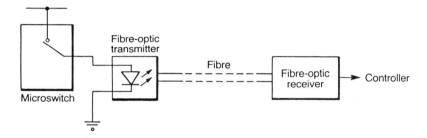

*Fig. 6.2* Fibre-optic signals over a long line.

Figure 6.2 shows a simple fibre-optic system. The microswitch could, in this case, be a simple on/off switch mounted near to the fibre-optic transmitter. It has to feed a fairly high current into the LED in the transmitter, which would be difficult to derive purely from noise – the LED has a low impedance (to ground in this case). When the switch is made, light is generated and fed down the fibre to a receiver which is somewhat more complex than the transmitter. This detects the often very low level of light signal from the fibre, and converts it into an electronic signal for the controller. A fibre gives the ultimate in electronic isolation from noise and spurious voltages from the machine under control. No common ground line is required, for instance, between the system under observation and the controller electronics.

Fibre-optic light guides are usually fed from an LED emitting in the red and infra-red end of the spectrum, and fibres of a kilometre or so are perfectly possible with low cost systems. If there is not enough signal transmitted along a long line, it is a simple matter to place boosters along the line, though it must be remembered that these will need power sources, as does any electronic circuitry. Short lengths of light guide are made from plastic, but the longer lengths will be glass fibres, often made from two refractive indices of glass to ensure total internal reflection within the fibre, and thus minimise loss. To further enhance fibre-optic systems, transmitters and receivers will often communicate the pulses by modulating a carrier. The two states, 1 and 0, are transmitted by different frequencies of signal rather than by on and off light signals. This means that there is always a signal of some kind being transmitted, and it allows detection of a break in the fibre, or failures in the transmitter or receiver systems.

Although microswitches have been given as examples of sensor signals here, the same general principles apply to any other sensors which transmit switched signals. For instance, proximity detectors give such digital signals. Thermostats also often simply change the state of a switch as their main ouput. Once again, switched information, like digital data, is easier to transmit over long lines than unconverted analogue data.

## OPERATOR SIGNALS

If there are just one or two buttons to be operated by an operator, then their interfacing to long lines is identical to that described in the above section. If, however, there is a front panel of keys, for instance, on perhaps several machines on a shop floor, and these keys are to be monitored continuously from some central point, then different tactics must be adopted. There is one main approach, with a number of implementations, to this type of problem. It should come as no surprise to learn that this involves conversion into digital signals, though there are several ways to implement this. As this is an important process control problem, we will look at the electronics of a possible method.

One of the most economical methods is to use a standard keyboard encoder, and simply arrange for this chip to send its output along some long lines. This condenses the problem of scanning the many keyboard switches, as was described in Chapter 5. An electronic system as shown at the foot of Figure 5.6 is used to scan the keyboard locally, and the six bits shown sent along long lines using the techniques of Figure 6.1 or 6.2 in the present chapter. If there are several such keyboards scattered around the shop floor, this requires a fair degree of cabling, but will achieve the aim adequately, using 24 V levels for each line as if it were a normal switched line. If visual feedback of the keypressings is also needed, the system becomes more complex, and the cabling needs increasing.

A better method of communicating the data of key-closure, however, is to digitise the data, and send it serially along a line. A popular protocol for achieving this is called RS232, and will be described in more detail later. For now, solving the problem of serially communicating the data from the keyboard will be looked at as an introduction to serial communications.

The principle behind serial communications is to take a binary number, split it into its separate bits, and send these bits along a single electronic line one at a time, in order — usually least significant bit first. Figure 6.3 shows an electronic system for this purpose. The keyboard is scanned by a

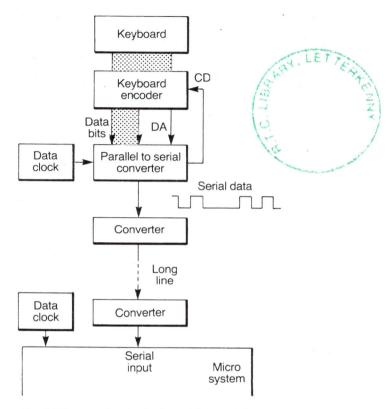

*Fig. 6.3* Communicating keyboard data serially.

keyboard encoder, exactly as in Figure 5.6 of the last chapter, and using the same control lines – DA and CD. The number of data bits is irrelevant, and for a large keyboard will be more than four bits. The keyboard encoder then feeds the parallel binary data into a parallel to serial converter which simply breaks it into separate bits, and sends them out serially to produce a square waveform as shown. This serial data is then converted into a high voltage or into fibre-optic transmission as appropriate. At the other end of the line, the data is reconverted and accepted by a special serial input in the micro system. The frequency of serial bit transmission depends upon the frequency of a data clock which produces an accurate square-wave input to the parallel to serial converter.

The micro also has a clock of the same frequency which it uses to reconvert the serial data into parallel data. Other more complex functions are also performed by the serial input device such as noting that some data has just been received, and perhaps alerting the micro to this fact. It is then up to the micro to accept the data and use it quickly enough that it does not miss further key-pressings on this or any other keyboard. The exact activity of running this system depends upon the software and upon the intelligence of the serial input device.

At the keyboard end, it is suggested here that the DA signal is not sent to the micro as described in Chapter 5. Instead, it is used to start the serial transmission process in the parallel to serial converter. When the data has been successfully sent, the parallel to serial converter signals along the CD line to clear DA without the micro being involved. This is a simplification of the keyboard side of the system, and would probably require a few more chips than implied by Figure 6.3. Also, the problem of whether the data clocks at the two ends of the long line have to be in phase, or just of approximately equal frequency, is vital. This is partly explained by the section below on serial communications. However, the full details are left to more complete texts on the subject of computer electronics. The reader is directed to the Bibliography for more detailed information.

The problem of returning information to the operator may also be solved by using non-MPU-based devices such as are shown in Figure 6.3. However, it is often decided that a minimal micro system is far more suited to these rather complex tasks than such 'discrete' systems. If a project is encountered where numeric or other operator data must be sent to and from a node, it will often best be performed by using simple SBCs in the nodes. Instead of designing more and more complex discrete electronic logic systems, the whole problem can be off-loaded to a standard micro system and some software. The micro is thus viewed as the most general type of electronic logic circuit.

## COMPUTER DATA COMMUNICATIONS

As can be seen, even a simple process control system may need the use of micros in the nodes on a shop floor, and there are many other cases where data must be communicated in systems both large and small. For instance, when a computer needs to print out a report, this text data must be communicated to the printer byte by byte, or bit by bit. In fact, this is the most common application of communications in computerised equipment. Another important application is for the sending of more complex data such as analogue signals, or 'stepped' data such as the position of a many position rotary switch. In each case, the data can be turned into binary numbers and communicated along the line.

We will now look at two methods of communicating this type of data in order to see a comparison between the two most popular communication methods. One is a parallel system and the other serial.

## Parallel communications

One of the objects of communication between machines is to transfer computer-based information in a file on one machine to computer-based information in a file on another machine. Of course, not all communication is meant to transfer between files, and the recipient may, as mentioned above, be a printer or other form of simple machine which uses the data immediately for some purpose. The full electronic details of the transfer are beyond the present scope, but this is such an important area that the main concepts are described here.

Let us assume for the present purposes that data is generally held in 8-bit memory locations within the computer. This may be purely binary data perhaps mirroring the individual bits of some I/O ports, or it may be text using a standard code such as ASCII – short for American Standard Code for Information Interchange. This latter provides, classically, a 7-bit code for each upper and lower case letter, number, punctuation mark, special character, as well as a number of codes for particular commands. The seven bits are generally packed out to eight bits by the addition of a leading zero, and thus fill a typical byte-wide memory location exactly. Different machines deal with the 'redundant' eighth bit in different ways, but this is not relevant here.

Figure 6.4 shows a sketch of the process of parallel communications. This shows a business computer communicating to a receiving machine. The file of text to be transferred is stored in the form of 8-bit bytes, and so it is logical to try to transfer eight bits at a time, for maximum efficiency, and indeed this makes for very fast transfer of data. The file is fed, byte by byte into an I/O chip, and some communications software makes sure that the data is presented only as fast as it can be used by the receiving device. The sender and the receiver are connected together through a ribbon cable, with more than eight lines to include extra lines for control and a common electronic ground.

The actual process of transfer is regulated by some special lines called handshaking lines. The sender (Tx) places a byte on the ribbon cable, via the I/O chip, and then signals the presence of this new byte to the receiver ($R_x$) by changing the state of the 'strobe' handshaking line. The receiver notes this state, when it comes round to needing the next piece of data, reads the byte, and when it is satisfied with the byte, it signals that it wants a new byte by changing the state of the 'busy' handshaking line. The sender monitors this line, and when it sees that the next byte is required, it resets the strobe line, to show that it is fetching the next byte, and that the one that is on the ribbon is to be disregarded. When it has the new byte set up on the ribbon, it again signals along the strobe, and the whole process repeats. The receiver keeps its busy line in the busy state until it is ready for the next byte.

As the handshaking control above is sequential, there is no danger of the

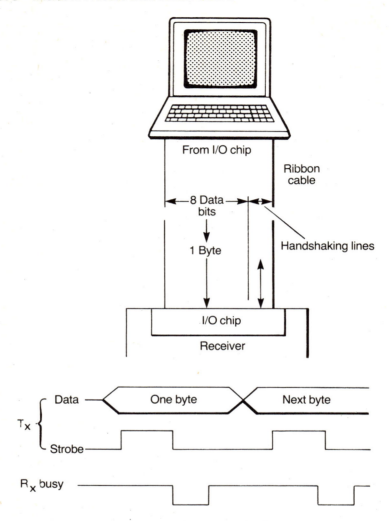

*Fig. 6.4* Parallel communications.

sender and receiver becoming out of sync. The data is transferred in an orderly and safe manner. However, despite the work that is necessary during the process, the transfer is very fast, and well suited to the sharing of large quantities of data at high speed.

The most common parallel I/O standard is the 'Centronics' standard; it is well defined, and most Centronics I/O ports will interface with very little trouble. As computers are themselves parallel devices, internally, parallel communications is very well suited to their general nature.

Printers generally have a parallel input port of the Centronics standard, and there is rarely any trouble with their connection. It is rare, however, for a computer to allow both input and output via its Centronics port, as most intercomputer communications are via serial ports. The serial port is the longest established method for general data transfer, and this is dealt with in the next section.

Centronics, and most parallel I/O processes, occur at the normal +5 V logic levels, and thus the length of a parallel cable can turn out to be rather restricted if the highest speeds of transfer are attempted. This is less of a problem with serial. Also, of course, laying parallel cable is more costly than serial, and parallel I/O cannot be effected via the telephone networks. These are the prime reasons for the serial I/O standard being adopted for general data transfer. However, it is possible to extend parallel cables over a long distance with appropriate electronics, but it is expensive and cumbersome in terms of cabling. Also, it is a general principle that the longer the cables, the more difficult it is to maintain the high speed of transfer.

Another parallel communications system is called the GPIB (general purpose interface bus) or IEEE interface. This was originally developed to allow a standardised method of communicating data between laboratory and other electronic instruments. The underlying physical principles are the same as for Centronics, but the exact details are different. Again, the idea is to define a standard which allows devices to plug directly into each other, but in this case the bus allows several devices to be plugged in it at the same time. It acts similarly to a data bus with several chips connected to it. There is a method of activating just one device at a time to communicate to the GPIB to prevent electronic clashes from occurring.

## Serial communications

There have been many different standards of serial I/O and this subsection describes the currently most popular. It is commonly known everywhere as RS232. This is a little inaccurate as the original definition of RS232 was for a set of control lines, and a definition of the electronic levels on the serial lines. The exact make-up of the the binary data to be sent, and its form were not defined.

As mentioned above, serial communications depend upon taking the 8-bit bytes, converting to individual bits, and sending these down a single line, in a serial stream, in the order in which they appear in the byte. The LSI I/O chip called a UAR/T has already been mentioned as a typical device for both splitting the bytes up into a serial stream, and reassembling them at the other end. There are other types of serial communication which are affected by other types of chip, but we are dealing here with what is known as asynchronous communications, and thus the UAR/T is the chip to use.

Figure 6.5 sketches some of the details of asynchronous serial

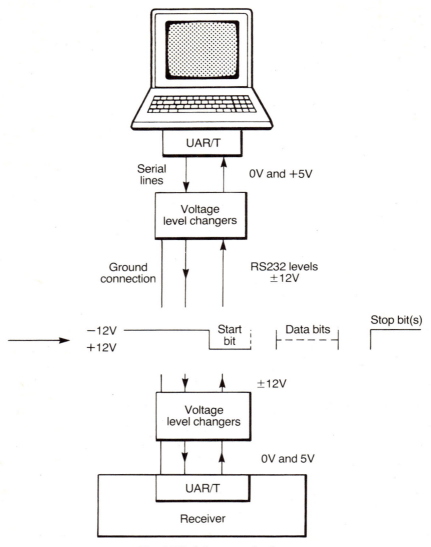

*Fig. 6.5* Serial communications.

communications. The popular use of the term 'RS232' (or RS232C to be more exact) includes the data format shown in the diagram. Again, a file of bytes is converted into a bit-serial form, and sent along a line to the receiver. There, the serial stream is reassembled into byte form for normal file storage.

The simple statement of the process above begs a number of questions. First, how is the receiver to know where one serial byte ends and the next starts? Second, how will the two ends of the line synchronise themselves so that they agree on the bit time, or the length of time one bit occupies in the transmission. If this is not clearly understood, then in the case where the sender sends two or more bits which are the same, next to each other, the receiver will not know where one starts, and the other finishes.

For these, and other, reasons the actual form of the serial data is rigidly defined, so that different machines can fully agree, and not become confused. The definition which is generally agreed is shown in the diagram. Each byte sent is framed around by a start bit which is always a logic zero, and at least one stop bit which is a logic one. In fact, the logic one is converted, in RS232, to −12 V on the actual line, and the logic zero to +12 V. When the receiver sees a zero on its input line, from scratch, it knows a data byte is to follow. It then counts the bits, as it captures them, and knows when the stop bit should arrive, as it has an internal clock which is set to the same rate as that of the sender. Sender and receiver have to agree on this clock rate before communicating. The bit rate, in bits per second, is called the 'baud' rate, and thus each machine must be set to the same baud rate before commencing transmission.

If the receiver finds no stop bit, it signals an error. Also, in some cases, an extra 'parity' bit is sent to allow for further error checking. Furthermore, it is not uncommon for further error checking data to be sent within the serial stream, before and after each given block of data. The problem is, how does one stop the transmitter, or make it repeat a block, if an error is found? Similarly, the receiver has a lot of work to do during receipt. It has to recognise a start bit, count the bits, check for stop bits, possibly for parity, assemble the serial byte into a parallel byte and then store the byte in a file. It may be that the byte is to be stored to disk, which is a comparatively slow business. As you can see, just as for parallel communications, some handshaking is essential.

Handshaking in serial communications can take one of two forms. There are some standard electronic lines which can be connected between the computers, along with the transmission lines and a ground connection; these act very much as the strobe and busy lines of Centronics. Of course, this increases the number of wires, and cannot be connected through the telephone system. However, it provides the fastest form of handshaking in RS232 transmission.

Another method of handshaking is called 'software' handshaking. Again, the system relies on a sequence of operations, and it requires communication lines in both directions. However, this only requires two communication lines, and a common ground connection, giving three wires, unless fibre-optics are used, where two fibres are required. It is possible to send serial data through the telephone system using software handshaking. The receiver sends a special control code to the transmitter, called an 'XOFF' character, when it wishes the transmitter to stop to allow the receiver to finish some action. When the receiver is ready for more, it sends an 'XON' character, and the transmitter continues. The transmitter must monitor its UAR/T continuously to look for XOFF and XON characters, and this takes time.

The fastest practical RS232 communications occur at around 2000 characters per second, or 20 000 bits per second, although there are plenty of special high speed serial communication standards which occur at many times this speed. The standard maximum length of an RS232 cable is around 15 to 20 m, though by using special drivers this may be increased. Other standards and electronic definitions such as RS422 and RS423 allow up to a mega bit per second, and distances of 1500 m to be achieved. In general, if high speed and long distances are required, the current cost of fibre-optics, along with its advantages make this a very viable alternative.

This was found to be the case in one particular extended data collection and production control system for an apparently secondary reason. The safety regulations governing conductive cable meant that data cables could not be run in the existing conduit system next to high voltage wires, and that special conduit would be required. It was found considerably cheaper to run long lengths of double-fibre light guides throughout the factory, and literally hung around any convenient strut within the building, as they are inert and non-conductive.

*Serial handshaking*
In general, the most easily interfaced RS232 standard is the software handshaking system; it rarely goes wrong. Hardware handshake can be a little more difficult to set up. It is often said that there is no proper standard for RS232, and that it is difficult to interface by this method. However, if you stick to software handshaking, and three wire communication, there is little to go wrong, apart from disagreement on the baud rate at each end of the line.

This describes RS232C which occurs using +12 V and −12 V along the line. Another method uses a current loop whereby the data is sent by pulses of 20 mA, again to defeat the noise pickup of a long line. In fact, historically, this was the original method of communicating serial data, and will be found on old teletype printers.

# Chapter Seven
# Further Concepts and Examples

## INTRODUCTION

The previous chapters have provided the 'nuts and bolts' of a variety of hardware and software control concepts. Using the data contained there, you should be able to view a possible product or process control project with a certain degree of familiarity, and it should enable an overall system design to be attempted, with the actual details being left for a specialist. The theory of control, along with the mathematical analyses of control systems, has been left for other books on the subject. In general if a full mathematical analysis is required, this would set the overall philosophy of the system, and impact the software and function of the system rather more than the choice of controller, which is the main subject dealt with in this book.

As an adjunct to the main topics introduced in the preceding chapter, it is instructive to look at some general software concepts. To this end, we will look briefly at the software hierarchy which is found in computer systems. This will explain some of the jargon words which are found in general use.

The next part of the chapter gives some generalised examples of systems which have been researched and proposed, or are actually in current use. These examples are not necessarily described in their real form; they are to be used as the basis of further ideas in your own industry.

## SOFTWARE HIERARCHY

This is not the place to start describing programming, nor to compare different computer languages. There are many excellent books on the various languages which may be used, as well as books on the theory of constructing efficient programs, and the philosophy of control theory in general. Instead, as a final component in the description of the internals of control systems, we will look briefly at some of the jargon which is

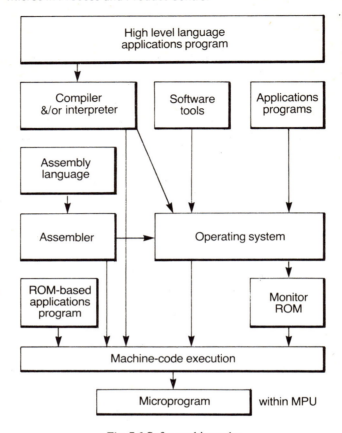

*Fig. 7.1* Software hierarchy.

encountered in the software side of a project. This does not involve a look at programming, but rather a look at the overall view of software itself.

Figure 7.1 shows a sketch of some of the main software concepts which will be met in this field. The MPU itself understands a language called 'machine code'. This consists of commands to the MPU such as ADD, STORE, READ and so on. Each of these commands is assigned a given binary number, determined by the designers of the MPU, and not necessarily following any logical recipe. Each binary coded instruction which is fetched by the MPU is executed by the MPU's internal 'microprogram'. The MPU itself, therefore, actually contains a program of its own, which is to be regarded as at the lowest level of the software hierarchy. This microprogram is not, in general, accessible to the user of the MPU; he simply accepts its actions as fixed, and uses the MPU accordingly.

The next level up in the hierarchy is the set of machine code instructions held in RAM or ROM within the system's electronics. This may take the form, for instance, of an EPROM holding a controlling applications program which has been developed elsewhere, blown into EPROM and plugged into place in the controller. In fact, this is quite enough software for a controller, and none of the rest of the diagram need apply to such a device.

If the system shown is in fact a larger computer system, it may also have a special 'monitor ROM' which contains a non-volatile program which takes over when the system is first switched on. It may also perform some of the low level housekeeping tasks such as running the keyboard and VDU screen, or some background interrupt routines. It is also in machine code.

If the system is a computer, it will probably be running an operating system (OS) which contains such routines as a disk operating system (DOS) and various routines for running the keyboard and VDU screen, if these are not handled by the monitor ROM. The operating system acts as a general operator and peripheral interface, and there are many such operating systems in existence. Sometimes a computer can change its OS, though often it is fixed when you buy a system. Certain operating systems are better at some operator interface tasks than others. In addition, some standard software will only run on a selected set of OSs. It may then be necessary, if possible, to change the OS of a given machine to cater for such software. However, the move these days is for most software package writers to make their work available on most serious OSs. The OS is supplied in machine code, and thus sits close to the lowest level of the machine.

All computer systems will be able to run special software tools for the development of programs, as well as standard applications programs such as word processors and accounting systems. These packages add to the facilities of the OS, and as such work through them, though they are also in machine code. The reason for a package running through an OS is that the writer of the package does not then have to concern himself with duplicating the important operator interfaces, or the interfaces to the disks and printers, etc. All that software should already be in the OS. Sometimes, it is in the monitor ROM, but in general it is better to allow it to reside in the OS as there is more room available in the OS for complex software modules. Also, this approach allows the flexibility of changing the OS and all the interface facilities in one go.

If a new program is to be written, it can be written directly in machine code, using the given binary number codes for the MPU's instructions. However, though this produces fast and memory efficient programming (code), it is very time wasteful. If machine code writing is needed, it is better to write in the next level of language up – namely 'assembler', or 'assembly language'.

Assembler uses mnemonics, or shortened forms of the actual machine code instructions. For instance, addition might use the mnemonic 'ADD', which is considerably easier to write and remember than '00101001', or whatever the binary may be. In order to turn these assembly language instructions into the binary equivalents, known as 'object code', a special program called an 'assembler' is used. The assembler is supplied in machine code, and would normally also work through the interfacing modules of the OS. When any program package is purchased, it is normally specified for a given OS, or at least a given computer system if there is only one OS.

The easiest languages to use are the 'high level' languages such as BASIC, PASCAL, COBOL, FORTRAN, and so on. These languages allow the programmer to write his instructions in near-English, and this enhances his ability to grasp the language, and write his program instructions at high speed. Such languages have to be turned into object code (machine code) so that the microprogram can understand the instructions. This requires a special program called an 'interpreter' or 'compiler'. Such packages again work through the OS, for their operator and peripheral interfacing. The object of these packages is simply that of translation. This can be a complex job, but is on a very different level from that of understanding natural human language which is full of nuance and connotation. Computer languages are exact, and not open to interpretation.

When a high level language is used to produce a program, it is tested, 'debugged' and probably finally converted into machine code using a compiler. At this point it becomes another applications program working through the OS.

Programs for controllers are often written in assembler so that the programmer can be as close as possible to the actual hardware itself. By writing in near binary in this manner, he can influence the I/O lines and memory contents in a more direct manner.

Applications programs often reside on disk, and can be called up at a minute's notice. This allows the general computing machine to act as a wordprocessor, an accountant, a program development system, or any other device for which a program has been written. This is the power of the stored program computer which Charles Babbage envisaged, but which required microelectronics for its implementation.

## SOME EXAMPLES

The rest of this chapter is composed of examples of control systems, and these range across the spectrum of the possible applications for microprocessors. We will examine large and small systems, in brief, and bring out the important aspects of the designs which underline the concepts introduced in this book.

Once again, rather than describe technical detail, the level of explanation is similar to that of the previous work, and simply points the reader towards the ideas of the control solutions.

### Food process industry

The food process industry supplies many examples of places to use microelectronics. Many food industries are very labour intensive, and this makes the introduction of automation and control particularly difficult. Some food industries, however, are already highly automated, and the use of microelectronics within aspects of these operations is simple and obvious, and slots in well with existing equipment, even though some of it is thoroughly pre-microelectronic in nature.

For instance, an old process line which manufactures biscuits, and has done for tens of years using the same mechanical equipment, is amenable to a number of microelectronic control devices which can make it more efficient and save money. For instance, as has been mentioned before, the problem of recognising waste biscuits without employing expensive labour can be solved using simple camera techniques. The extra devices to be added to the line, no matter how old, would be a sensing station, an air blower and a waste shute for the damaged biscuits. Most process lines will be able to find some place to retrofit such equipment. This can save labour in terms of inspection before packaging, and is highly cost-effective within a short time. It can, of course, be applied to a number of related processes in other industries too, including, perhaps, the manufacture of nuts and bolts, or the press-cutting of gaskets.

Another use of microelectronics, which is more and more important in industry, is the replacement of large relay-based controllers by microelectronics, sometimes at a fraction of the cost of the original equipment. The advantages are greater reliability in some cases, but much more important is the remarkable increase in the flexibility of control. Over the years such relay equipment will have begun to wear out, and the users will have discovered more and more facilities which they wished it had, but which are too expensive and involved to add to the obsolete controllers.

However, the real opportunity for control engineering is in the more labour intensive parts of food processing. This area has not begun to be considered properly in the industry, and has yet to be taken seriously by the majority of manufacturers. A typical example may be a large processing hall where meat products are issued at one end of the hall, and pass down lines of people cutting, removing waste, weighing, collecting into groups, packaging, labelling, and finally packing into cartons to be stored in the warehouse. These operations may be completely manual, and will probably remain so for the forseeable future. To see why this is, consider the problem of taking a large cured pig carcase, and turning it into joints and various cuts of meat. This is such a human-orientated decision process that its automation by robotic equipment would be a major feat of pattern recognition, let alone automated handling. This is certainly not the main place to look for introducing the advantages of automation and control – at least initially.

The correct place to look is in the control of the production itself. In just such a case, a successful production control system was put in which controlled all the activities of the production hall automatically, and could produce complete management reports on the process at any instant. This latter was particularly important as, previously, somewhat inaccurate reports were only available several days after the event – too late to correct some problem before it had caused a serious setback to the whole of production.

The method of control consisted of a hierarchy of computers, starting with a high level network designed to aid the production planners in constructing a weekly and daily plan dependent upon current orders. This plan would then be transmitted down to the production hall, using fibre-optic cables. On the shop floor were placed several intelligent nodes which could issue instructions of great detail to the operators. For instance, one was placed at the point where raw material was issued to the production hall, and this node weighed the incoming meat packages and ordered these to be taken to a particular line to be dealt with accordingly. During the process, data was continuously collected, and special automatic labelling equipment used to ensure that the system closed the loop fully between planners, production, and the warehouse itself.

Each node was capable of being given a plan, and even working alone for several hours if the central computer should fail. When convenient, the central computer network would request current production data from the nodes, including one which collected data on waste from the process. This data could then be used to compare the daily and weekly plan with the current position. In this way, the planners could instantly pick up any problems, and the current state of the production hall, particularly yields, was known, and fully controlled, at any instant.

This is an example of closed-loop feedback through a weld of humans and machines. This example takes the idea of data collection to its logical conclusion. Here the machinery is issuing instructions, checking on progress, and reporting back to the production planners.

Another example of a system for a labour intensive process is that of type and weight grading. For instance, animal carcases may be passed into a production hall where they are graded by type and weight, and dropped into bins, accordingly, for later packaging and labelling. The following describes the elements of a system which was proposed to a particular manufacturer.

In this case, there are two classes of carcase, and around eight weight ranges for each class. This makes sixteen bins. The classical method is for a load of carcases to arrive at the grading hall and be placed on large tables. Operators then class and weigh them by hand, and place them in the appropriate bins. This is very labour intensive, and an excellent candidate for automation. The consequence of this is either a reduction in the labour force, or a dramatic increase in productivity for a given level of staffing.

We will now look at a proposed electronic controller solution to the grading problem. However, in order to even start such a system, the mechanics must be designed to be amenable to such a solution – automating a bunch of tables piled high with carcases is not an efficient route to take! The first problem to solve is that of the transport of the carcases into and through the system, and possibly their transport out of it. In this case, the carcases were already on an overhead conveyor system, and this could simply be extended to bring the product into the control environment. The object of the system would be to place the carcases in bins as before. The transport of the filled bins was considered to be an external problem.

It is typical of the line of attack in such a process to start with the mechanics and try to make this more linear and streamlined. The more physical work which can be performed by simple mechanics, the better and cheaper the final system becomes. In this case, as we shall see, the use of a conveyor system simplifies the control to a considerable degree.

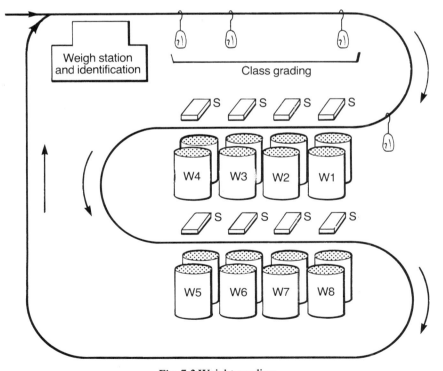

*Fig. 7.2* Weight grading.

Figure 7.2 shows the main elements of the system. A conveyor system is fed with carcases hanging on hooks. The first section of the conveyor has a weigh station which simply passes the carcases on their hooks over a load cell. At the same time, the hooks themselves are identified so that a given hook is identified from that moment on as holding a carcase of a given weight. This assumes that each hook has a unique identification on it which is machine readable. This is achievable in a number of ways. For instance, the hook could have a vertical metal tag with slots cut in it like a computer card. An optical reader could then read the code and thus identify that hook and its encumbent carcase throughout the rest of the transport system. This identification is important for later removal of the carcase into the correct bin.

The next stage in the process is for human operators, in the 'class grading' section, to decide to which of the two possible grades each weighed carcase belongs. The class is stored at this point in the mechanical system by turning the carcase one way or the other in some manner so that, when it is removed from the hook later, it falls one way or the other from the conveyor, into one of a pair of bins at each weight sorting stage W1 to W8.

When a given carcase reaches the weight sorting stages, its hook identification will be read by one of the blocks marked 'S'. These S blocks, along with the weigh station itself, will be tied into a central control system which we shall discuss shortly. For the moment, assume that this controller is checking the identification of each carcase at each S block until it ties up its memory of the weight on that hook with the fact that it has now reached its correct weight stage. At that point, the controller instructs a solenoid to push the carcase off the hook so that it drops into a bin. The hooks have already been turned one way or the other by the human graders, and the result is that when the hook releases its load, the carcase naturally drops into one bin or the other according to grade.

The alternative to this method of turning the hook might be for the human operators to drop a further grading code onto the hook, and have this read by the S blocks. The hook could then be instructed to despatch its load into one or other grade of bin, as well as the correct weight. When the hook has deposited its carcase, it returns to be filled again, and the system erases its memory of each hook as it drops its load.

The system is naturally self correcting if a hook does not deposit its load. The hook simply re-arrives at the start of the system and is treated exactly as it was before. If a hook arrives without a load, this is also detected, and the system simply disregards it throughout.

This arrangement can save a large amount of labour and thus expense, or allow a manufacturer to expand considerably with existing staff, and thus better assure its future. Either way, the system will probably be cost justifiable in a year or two.

The details of the design of the mechanics are beyond our present scope, but some advice is worth mentioning. In this system, for instance, each S block has to identify the current hook, and each hook has to be separately manufactured to be uniquely identified. This can be circumvented by using a counting system. For instance, every fifty hooks could be a uniquely identifiable dummy, perhaps simply by its bringing a vane in front of an opto-sensor. This would start the system counting hooks at the weigh station. Each hook would be identified with its position in the line of hooks, and thus associated with a weight. At the first S block, the dummy could be identified, and a simple counting procedure, using opto-sensors, or proximity sensors, at each S block would identify when a given hook and carcase had arrived. If there are several dummy hooks in the line, the system would be self resetting, and hence self correcting on a fairly short time scale. This also makes the problem of manufacturing the hook system somewhat simpler.

We will now look at the systems analysis of the controller, and choosing a suitable level of device. The first question is how many I/O lines are needed, and what is their character. To answer this, Figure 7.3 summarises

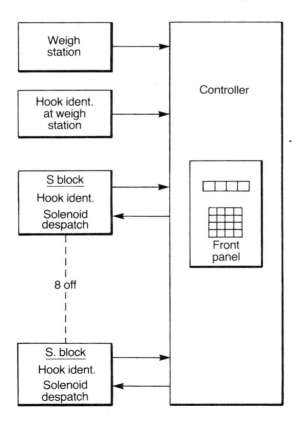

*Fig. 7.3* Weight grading controller.

the control requirements. The weigh station has a weighing sensor and a hook identification sensor. Each of the eight S blocks has a hook identification and a solenoid to despatch the carcases into bins. We will assume the class of carcase is entirely outside the controller's scope, as described above. The controller itself would probably require a little front panel information such as error lights. This would be a simple minimum system, and may require considerable expansion to allow for other functions within the grading hall. However, assuming that this system is adequate, we will look at the sort of controller characteristics which are required.

The number and type of I/O lines can be estimated before detailed design is started in the following manner. The weigh station is the first block to choose. It can either be a fully electronically interfaced device which outputs a digital signal along an RS232 line when ready with a weight, or it can have an analogue output, already interfaced to allow an ADC to connect to it directly, as described in a previous chapter. The alternative is that a bare load cell is purchased, and the controller would have to contain the appropriate analogue interface circuitry for conversion to digital values for the MPU system. In a physically extended system, it would be an advantage to provide the main part of the interface at the load cell itself, in order to send digital information over the long line to the controller. This could be via an RS232, or other more extensible serial protocol. This is an important part of the system, and the controller itself can only be fully specified after this component has been chosen. The choice may also be decided partly by the accuracy with which the weighing is to be effected, and the range of interfaces offered for a given manufacturer's weighing machines.

We will assume here that the decision is taken to purchase a weighing device which sends out an analogue voltage, and that the weigh cell is up to 300 m from the controller. This means that a voltage to frequency converter could be used to send pulses along the line to a single input line on the controller.

The next question is to consider the nine hook identification sensors in the complete system. We will assume that all sensors and solenoids in the system must work at a long distance, say up to 300 m, from the controller.

We will assume that each hook sensor consists of two separate sensing elements. The first element is simply a presence sensor. That is, it will detect the presence of any type of hook by a photo cell. The second element will detect the presence of the special dummy, of which there is assumed to be just one in this system. Thus each hook sensor has two photo cells.

An important function of the presence sensor at the weigh station is to allow the system to detect when a carcase is in the correct position to be weighed. The weigh sensor would be expected to be a little extended along the conveyor line, but it is important that the weight is read at the right time, when a caracase is in the best position on the weighing machine. It should also be remembered that as the hook rolls over the weigh station, a certain amount of mechanical bounce will occur which will not give a steady output. It is important to take a number of readings and average these out, perhaps discarding any which are wildly out of a given range. This is mostly down to the software. However, if a voltage to frequency converter is used, the frequencies of the pulses sent must be much higher than the characteristic frequency of the bouncing to ensure that the two frequencies do not interfere.

From the above it is clear that each sensor station has two lines back to the controller – one for hook presence, the other to identify the dummy. This makes a total of eighteen inputs to the controller. At each end of the long lines, appropriate interface circuitry must be included to ensure that electrical noise does not interfere with the signals. However, these are comparatively slow signals, widely spaced, and thus will not need particularly complex interfaces.

The solenoids are simply on/off devices which must be controlled to despatch carcases to the appropriate bins. This gives a total of eight outputs from the controller.

The front panel will be assumed to require a couple of lights and perhaps an audible warning device, although the complexity of the front panel is up to the user of the system, and the parameters he would like to monitor. We will assume a basic four lines of output from the controller for this front panel. If further peripherals such as printers and connection to other computer systems are required, this will increase the I/O lines, and the software of the system.

The total number of lines required in the simple system is thus nineteen inputs and twelve outputs. Some interfacing is required for all these I/O signals, as they are to be connected over long lines to the production line. Even the front panel will require some interfacing to drive the lights, etc.

The software required to run the system will have to ensure that no inputs are missed – all inputs could be simultaneous, but the time between inputs on a given line is comparatively large. This means that inputs on a given line should be latched at the controller, and read by the software in a polling loop which also clears each input signal as it is read, in order to ready it for the next input. By this means, no inputs should be lost. However, if there is any problem, it can probably be solved by using an interrupt driven polling loop which simply polls and stores the state of all inputs every 10 milliseconds, or something similar. If the inputs are not latched, care will have to be taken not to accept any given input more than once through multiple reads, just as was described in an earlier chapter for keyboard reading. The results of the reads can be stored in memory in some fashion, and used by the main program when it has time.

In general, the working of a system such as this is several orders slower than that of the MPU system, and there is rarely any time problem.

The type of controller needed here should now be clear from the criteria given in the first part of the book. A large computer system or a multi-board controller would seem to be unnecessary, and a simple SBC will give a cost-effective approach. However, if analysis and reports are to be given by the system, a small computer system may be appropriate with multiplexing on whatever I/O lines it has, in order to extend it to the number of I/O lines required here. The front panel could be replaced by the computer screen. However, it is often rather less cost-effective to produce production controllers using small computer systems, and in general a robust and easily developed SBC approach would be taken. Alternatively, in a sequencing operation of this sort, a PLC might be appropriate, although a rather complex one would be needed to sort out the logic of identifying the hooks by a counting method. Such software is usually better developed on a normal MPU-based system which has the flexibility of any amount of logical complexity in its function.

A major point to appreciate is that even though the system is extended over a large industrial area, the actual control of the process is straightforward, and the controller has no 'knowledge' of where its I/O signals are coming from or going to.

## The use of continuous weighing

In the above example, continuous weighing was used to decide on the controller's function with regard to each item in the production line. No long term note of weights was required, though this could be added to allow a certain amount of production analysis to be formed, though this data would either make the front panel more complex, or demand a link to an external computer system to display and use the result.

Another important facet of continuous weighing concerns recent legislation which allows a manufacturer to fill containers with product according to a statistical method, rather than an exact minimum volume or weight. For instance, in filling tubes will glue, if the stated quantity is 3 g, the fill can be less than this if, over a large number of fills, the average is 3 g. Of course, there is an allowable spread too, and too low a fill in any given container is still not tolerable. Weighing systems can be used for two purposes in this example.

The first purpose of a system might be purely to monitor an existing filling process, and ensure that the law is being adhered to without wastefully overfilling. In this case, any process adjustment is made manually. The second purpose might be to include the weight data within the control loop of the filling mechanism. The controller switches the fill head on until the correct weight is seen to be reached – this might require a completely new system to be designed.

Continuous weighing assemblies are not mechanically simple. It is essential that the load is correctly distributed on the weigh head. The load must be placed on the head and a certain amount of settling time allowed before a reading is taken. This is also true of a conveyor belt weigher, where a certain amount of time is allowed between the load arriving at the weigh area and a reading, or series of readings, being taken.

In the case of a controlled filler, the weigh sensor forms part of the feedback loop of the system. When the vessel is in place, the fill commences, and continues until the weight has reached the correct level. At this point, a fully automatic system stops the fill, moves that vessel off the fill station and puts the next vessel in position.

### Example of the use of a light sensor

There are applications in industry for optoelectronics which have already been described in detail. However, there are also applications for sensor systems which have to calculate the amount of light which is emitted or passed through an object. An industrial example of this is in the electroplating of small flat metal items. This is not really an automatic on-line control problem, but it shows how a particular type of optical sensing operation can be useful in manufacturing industry.

Typically, the problem is that a sheet of small metal parts for electrical contacts, for instance, is to be plated with gold, or some other noble metal. The parts will have been press-cut from a thin sheet of metal, and will still be attached together in the original sheets by small tags left during the cutting. The problem is to plate the right amount of gold onto the parts, whatever their shape and size. This is determined by the electrical current and time for which plating is in progress, and the surface area of the sheet to be plated. The first two parameters are easy to measure with great accuracy, but the surface area is difficult. A typical method is to place the sheet on a light table, and shine light through the sheet between the metal parts. The light is collected and a light intensity measurement is taken to determine how much is obscured by the sheet. This can be fed into a computer system, along with the thickness of gold to be plated. Either the operator is given instructions for the settings, or those settings can be measured and adjusted automatically using ADCs, DACs and appropriate power interfacing.

This type of system can be highly cost justifiable in a short time. It may make little difference, in a given application, whether the process coats 5 microns or 7 microns of gold onto a contact surface. The accurate process control of the plating can save the extra 2 microns, however, and thus become quickly cost-effective.

## EXAMPLE OF A SMALL PORTABLE CONTROLLER

As an example of a small portable control application, consider a product which a manufacturer wishes to produce for the automotive trade. The object is to design and manufacture a range of upward compatible hand held meters which will start with a voltage measuring device for the spark at the plugs. This would be expected to work from a battery, dry or rechargeable, and display the result digitally for each plug in turn. Later it is hoped to incorporate a low tension voltage measurement, and then a vacuum gauge, and so on. The whole device must look appealing as well as being highly robust and capable of taking the inevitable knocking about associated with car maintenance.

The whole device would, thus, be confined to a single hand-held enclosure, and have a socket at the back into which a set of leads could be attached. Initially, these could be, say, between six and fourteen leads, to include one for each plug and an earth and car battery lead. This would allow the user simply to remove the spark leads, and place the voltage sensor between the spark lead and the top of the plug. In this way, the engine would continue to work normally, but the sensor could monitor the voltage at the top of the plug directly.

The function of the device might be to allow the user to turn a rotary switch, or use a simple keyboard, to command the voltage of any given plug to be displayed. By this means, the voltage of the spark to each cylinder can be monitored while the engine is turning at any required speed.

As the device must look good and be easy to use, much thought will be given to the external design of the device. It must be large enough to be capable of supporting expansion, but small and neat enough to be held and operated easily. This requires an industrial designer, and is beyond the present scope.

We will assume that it is decided that the device has a small numeric keypad, and a few extra function keys, plus a 4-digit numeric display. The internal electronics of the system will have to be MPU-based for two reasons. Firstly, there is a certain amount of work to be done by the machine to provide a meaningful display, and to allow a keyboard to be operated. Secondly, the expansion of the system will certainly require some calculation to be performed later if not sooner. The exact place on the waveform of the spark to take as the stated spark voltage is also an important parameter, and this could be sorted out easily using software. The freedom to provide an intelligent and complex analysis of the engine will only come from providing comparatively intelligent electronics.

The next step is to decide on the volumes of manufacture which are to be committed to. In this case, we will assume that the final product is to be low in unit cost and high in volume. The result is expected to be a product which will be sold cheaply, and thus produce the volume of sales needed as a consequence.

The situation above fits a number of areas of manufacture of products, from toys to simple test instruments in many fields. The only difference which may occur is in the provision of the power supply. For many small control applications, batteries may have to be used. This entails designing with LCD displays and CMOS technology chips throughout to save power. The conclusions as to the type of control electronics required here will transfer to any other similar system.

Having decided on a high volume device, the usual approach would be to consider a single chip processor, and try to confine the maximum amount of electronics to this device. The electronics will have to have at least one ADC to sense the incoming voltage, and an 'analogue switch' to allow the ADC to be switched to any of the incoming signals. An analogue switch, as its name implies, allows analogue voltages to be blocked or passed through, rather than just digital signals. There are standard ICs for this purpose in the CMOS range.

The amount of I/O which is needed can be decided from Figures 7.4 and 7.5. Chapter 5 has dealt with the interfacing of displays and

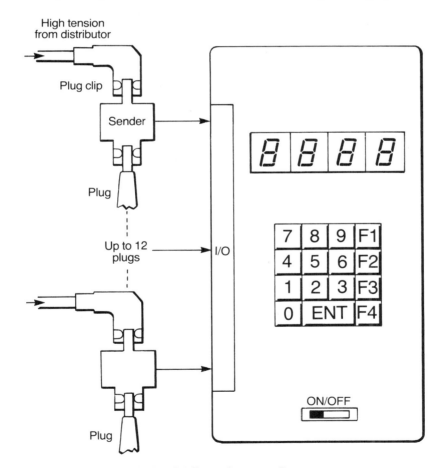

*Fig. 7.4* Car tuning controller.

keyboards, and the system would probably require a keyboard encoder and some ICs to run the display. Figures 5.3 and 5.6 show a typical system, and this implies that the I/O requirements would be of the order of twelve bits for keyboard and display. However, this can be reduced by including the keyboard encoder as part of the internal memory of the MPU by connecting it to the buses as shown in Figure 7.5.

The plugs would be connected as shown in Figure 7.4 through a voltage sender which would attenuate the voltage at the plug considerably before sending it down to the electronics. The sender would not need to affect the voltage through to the tops of the plugs noticeably. Up to twelve possible inputs should be allowed for, and a single ADC can be used to read any one of these inputs. The I/O would have to be able to switch a single signal through to the ADC out of the twelve inputs, and this requires four bits. This would, in fact, allow up to sixteen inputs to be monitored, and this could be useful for future expansions.

The main MPU could be a single chip device with a minimum of memory, and the program formed onto masked ROM. The alternative, and more expensive approach, is to use a single chip micro with external ROM, which could be EPROM in the first instance to allow easy development of the system. The single chip microcomputer will usually include some I/O, but if the buses are to be expanded to include other devices, this normally limits the amount available, so a further I/O chip may also be needed. However, given the system of Figure 7.5, just ten I/O bits are needed, and half a dozen modules, most of which are single ICs, though the display will have more unless a single chip LCD display decoder is used.

The single chip device only reduces the size of the system by a relatively small amount, and it may be decided in this case that the increased unit cost of having a conventional MPU-based system may be worth while to keep the development cost to a minimum. Either way, if all the components must be chosen to be low power, for a given application, this is possible with all the functions mentioned in this design. The MPU clock will be a single crystal, and it is useful to keep the frequency down to the minimum to save power. In general, the higher the frequency, the greater is the power consumption.

For cheapness and robustness, a customised membrane keyboard would probably be ordered for this project, and indeed this is a good way to affect the outside look of the complete device. Any style of graphics is available for the front cover of the membrane, and this can be used to provide an excellent and stylish finish to even the most functional of enclosures. The keyboard will need to allow the user to key in the plug which he wishes to monitor, and the function keys shown allow the user to command the device to show up plug voltage, or any future parameter which may be included.

Development would have to be performed using a large computer

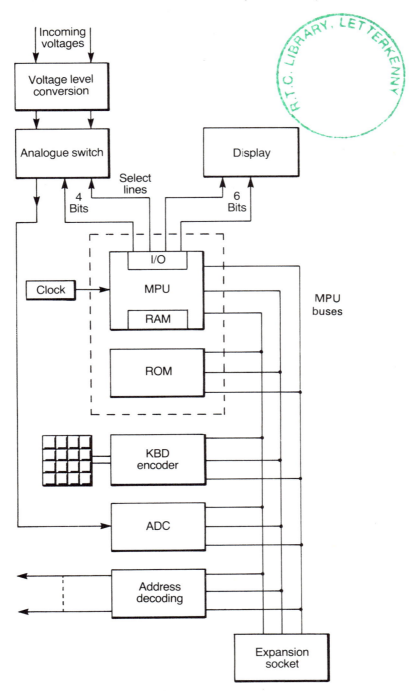

*Fig. 7.5* Car tuning controller schematic.

development system, and the alternatives for this part of the project have been described in Chapter 1.

An excellent method of producing this hand held device would be to develop the system using normal ICs and then, when working, have the complete assembly performed using surface mounted devices (SMDs). This keeps the prototyping straightforward and simple, but allows high volume techniques to be employed with little tooling and setup.

In many ways, the final controller chosen is dependent upon the number to be produced. For instance, if the volume were to be of the order of a thousand, the SMD approach would be appropriate. If a hundred thousand were to be produced, the amortisation of the development would allow far greater development costs. The correct approach in this case is to have the system produced using a single chip device, and order an additional special customised chip with all the other electronics contained on it, including, perhaps, keyboard encoder, display electronics, address decoding, and perhaps even the analogue switch and ADC. This keeps the assembly cost to a minimum, and produces a very low unit cost. However, development costs will be initially high.

## Appendix One
# Binary Numbers, Bits and Bytes

## INTRODUCTION

The text refers to a number of concepts in the field of binary numbers, and this appendix describes some of the background briefly.

## NUMBER BASES

We normally count numbers according to a scale of ten digits, which presumably originated from counting on digits of our two hands. This means that numbers are shown in the following form:

567384

This is shorthand for a number somewhat larger than half a million. The actual number labelled in this way is an underlying abstraction which we conveniently label as shown. However, there are many other ways to describe this number, and a familiar method is to use the Roman numbering system – this is left as an exercise for the reader!

The more topical method of labelling the number above is to use binary notation. In order to do this, we must try to understand the underlying principles of the above label, and then generalise to other systems. The number labelled as 567384 can be re stated as:

|   | | |
|---|---|---|
|   | 5 × 100000 | (hundred thousands) |
| + | 6 × 10000 | (ten thousands) |
| + | 7 × 1000 | (thousands) |
| + | 3 × 100 | (hundreds) |
| + | 8 × 10 | (tens) |
| + | 4 × 1 | (units) |

Each of the digits above represents a number of tens raised to some power. For instance, the 7 is a number of thousands, or 10 raised to the power 3. The last digit (4) is the number of units, or 10s raised to the power 0, because any number raised to the power 0 is 1. Thus, the labelling system for this number is to place the number of units to the extreme right hand side of the label, and the number of 10-to-the-fifths to the left hand side. The number of units is called the 'least significant digit' (LSD) and the digit at the other end of the number is called the 'most significant digit' (MSD).

10 is called the 'base' or 'radix' of our normal number labelling system, called the 'decimal' or 'denary' system. It is no coincidence that the number of digits required for the labelling, 10, is equal to the base. This is a rule for all numbering systems.

In computing, the 'binary' system is used, where 2 is the base, or radix, and the result is that just two digits are used – 1 and 0. These are called the 'binary digits' or 'bits' for short. The reason for the importance of the binary system is that modern digital logic uses components which are simple switches and thus either 'on' or 'off' or at 1 or 0 levels. A binary number would look like:

    10100101110110

Again, the digit (or bit) to the far left is called the most significant bit (MSB) and the digit to the right the least significant bit (LSB). The number label above is rather difficult for humans to comprehend, and it is often split into groups of four bits, starting at the LSB end, and adding any leading zeros which may be needed to pack the left hand side out to four bits. This is shown below:

    0010   1101   0111   1110
    MSB                  LSB

This helps with the writing of such numbers, but it also supplies a more convenient numbering system called 'hexadecimal', as we shall see.

The conversion of binary numbers to and from decimal is a little involved arithmetically, because the base of the binary system is a power of 2, and that of decimal is not. However, hexadecimal is based on 16, which is a power of 2, and conversion turns out to be very easy. Before we look at conversion in general, the following gives the sixteen digits of the hexadecimal system, with equivalent binary and decimal numbers. The binary numbers are arranged as 4-bit numbers for convenience by the addition of some extra leading zeroes which have no arithmetic effect, but help the later conversion considerably.

| Hexadecimal | Binary | Decimal |
|---|---|---|
| 0 | 0000 | 0 |
| 1 | 0001 | 1 |
| 2 | 0010 | 2 |
| 3 | 0011 | 3 |
| 4 | 0100 | 4 |
| 5 | 0101 | 5 |
| 6 | 0110 | 6 |
| 7 | 0111 | 7 |
| 8 | 1000 | 8 |

| Hexadecimal | Binary | Decimal |
|---|---|---|
| 9 | 1001 | 9 |
| A | 1010 | 10 |
| B | 1011 | 11 |
| C | 1100 | 12 |
| D | 1101 | 13 |
| E | 1110 | 14 |
| F | 1111 | 15 |

Note that the first ten digits of hex are the same as decimal, but then there are still six more digits to find to produce a total of sixteen. It was decided that the first six letters of the alphabet would be as good as any to use.

Note that the binary number given above, which we grouped into 4-bit sections, can now be converted to hexadecimal group by group from the above table. This gives the number:

2D7E

which is considerably easier for us to handle, but still represents the same abstract number. The decimal equivalent of this number is:

11646

## CONVERSION

To convert binary numbers to decimal, remember how the very first decimal number mentioned above was split into powers of the base (10). The same is true of binary numbers. For instance, the number 10110 can be written as:

$$1 \times 2^4 \quad + \quad 0 \times 2^3 \quad + \quad 1 \times 2^2 \quad + \quad 1 \times 2^1 \quad + \quad 0 \times 2^0$$

or:

$$1 \times 16 \quad + \quad 0 \times 8 \quad + \quad 1 \times 4 \quad + \quad 1 \times 2 \quad + \quad 0 \times 1$$

or:

$$16 \quad + \quad 0 \quad + \quad 4 \quad + \quad 2 \quad + \quad 0$$

The result is 22 in decimal.

If the binary number is to be converted into hex, it must be represented as groups of four. This is fine for the least significant group, which is 0110, or 6, but will need some extra leading zeros for the next group which thus becomes 0001, which is 1 in any base. The conversion is thus:

10110 = 0001 0110 = 16 (Hex)

But this shows the danger of simply writing down a number in some base without indicating the base being used. This number looks like 'sixteen', but is actually 'twenty-two'!

A possible way to write this number in the three bases here is:

$$(10110)_2 \;=\; (16)_{16} \;=\; (22)_{10}$$

Different computer languages have other methods of labelling their numbering systems, and you will have to read the manuals when you meet them. The essential point is that if more than one base is being used, confusion can reign if care is not taken. For instance, given the above, how would you know what abstract number is being referred to by the label '100'? Is it binary, hex or decimal?

To convert hex to decimal, the same principles hold, for instance, consider:

$$
\begin{aligned}
(A3E5)_{16} &= A \times 16^3 \;+\; 3 \times 16^2 \;+\; E \times 16^1 \;+\; 5 \times 16^0 \\
&= 10 \times 4096 \;+\; 3 \times 256 \;+\; 15 \times 16 \;+\; 5 \times 1 \\
&= \quad 40\,960 \;+\; 768 \;+\; 240 \;+\; 5 \\
&= \quad (41\,973)_{10}
\end{aligned}
$$

The conversion the other way for decimal to binary and hex is a process of division, and can be found in any standard book on the subject, including some of the books indicated in the Bibliography.

## CONSEQUENCES OF BINARY NOTATION

As mentioned in this book, the number of binary patterns possible on a given set of lines is an important number. For instance, if a chip has sixteen internal registers, it is important to be able to calculate how many address lines are needed to address it. Also, if an I/O chip has eight output lines, it is important to be able to calculate how many different patterns of 1s and 0s can be output on those eight lines in parallel. The answer is simple. $N$ lines will have 2 to the power $N$ different possible binary patterns on them. To see this, consider that one line has two possible states, (1 and 0) or 2 to the power 1 states. Two lines have four states (00 01 10 11) which is 2 to the power 2, and so on. Each time a line is added, it multiplies the number of possible states by two.

For instance, if a chip has four address lines, it can have at most 2 to the power 4 (16) different addresses in binary on those lines. A chip having eight outputs can have any one of 256 binary patterns of 1s and 0s on those lines at any instant. To select one of eight different peripheral devices, three address lines (with some electronics) will be needed.

Another important consequence of using binary is that binary numbers have to be used to represent all the symbols which we use in daily life. For instance, all the letters, punctuation marks, special characters, and so on. The numbers, of course, should be easy to represent using the base 2 equivalents. However, in order to represent all common human symbols, including the ten digits, there are some internationally agreed codes which match 7-bit binary numbers up with these characters. One of these, and the most common today, is ASCII, or American Standard Code for Information Interchange. This simply assigns binary numbers, almost at random, to the symbols. For instance, the following are some of the ASCII codes:

'3' is represented by 0110011
'A' is represented by 1000001

Note that these are 7-bit codes, and by increasing to 8-bits, and grouping into 4-bit pieces, they can be converted into hex as follows:

'3' = 0011 0011 = 33 (Hex)
'A' = 0100 0001 = 41 (Hex)

As there are seven bits, there are 2 to the power 7 different binary numbers possible, and hence 128 different symbols are represented.

Since the most common basic unit of memory is eight bits, this means that there is one wasted bit (the MSB or 'top' bit) in each location, and some systems use this to represent some extra attribute of the display of the symbol on a VDU screen. For instance, the top bit may be used to switch the symbol into reverse video whereby it becomes dark on a light background instead of light on a dark background.

The number of address lines given out by an MPU determines its maximum memory size. For instance, sixteen address lines gives 2 to the power 16 different possible addresses, which equals 65 536.

.

## SOME JARGON

Memory is normally split into 8-bit pieces which are called 'bytes'. A 4-bit group is called a 'nybble', and a 16 bit group, sometimes, a 'word', though this term is used in an ambiguous way, and should be approached with care. Also, as 2 to the power 10 is equal to 1024, this is a convenient unit to use to measure memory capacity in something approaching thousands. Thus, this unit is called the 'K'. The 65 536 address locations mentioned above are, therefore, 64K of memory, and a megabyte (MB) is 1024K. Thus, 1 MB is actually 1 048 576 bytes, and is a little more than a million bytes of memory. However, these numbers, K and MB, are sufficiently near to thousand and million for us to think of them in these terms.

A 16-bit MPU will allow much more memory to be addressed, by using more address lines. For instance, if 20 address lines are offered, raising 2 to this power gives 1 MB of memory, which is directly addressable.

This gives some of the main ideas needed to understand this book. If you are interested in more detail, refer to one of the micro design books mentioned in the Bibliography.

# Appendix Two
# Company Addresses

Arcom Control Systems Ltd,
Unit 8, Clifton Rd,
Cambridge. CB1 4WH
Tel (0223) 242224 Telex 817114 (Attn Arcom)

Arcom manufacture a series of controllers and control systems, including an SBC which has a resident BASIC interpreter and on-board EPROM programmer.

J.P. Designs
37 Oyster Row,
Cambridge. CB5 8LJ
Tel (0223) 322234

J.P. Designs produce a range of very low cost SBCs based around a number of 8-bit processors. In addition, they have a number of more sophisticated SBCs for larger applications.

Darenth Automation
132 Percival Rd.,
Enfield,
Middx. EN1 1QV
Tel 01-363-4476 & 367-0098

Darenth distribute the 'Beblec' PLC which has eight inputs and eight outputs, and is programmed directly by using pins in a patch board on the front panel.

Mitsubishi Electric UK Ltd,
Hertford Place,
Maple Cross,
Rickmansworth,
Herts WD3 2BJ
Tel. (0923) 770000 Telex 91656

Mitsubishi make some of the most up-to-date and well constructed PLCs on the market. They are fully programmable and start from a simple sequencer, and finish at the top end of the market.

Allen-Bradley Ltd,
Denbigh Rd,
Bletchley,
Milton Keynes. MK1 1EP
Tel. (0908) 71144 Telex 82396

Allen Bradley are a well established manufacturer of large PLCs.

Commotion
241 Green St,
Enfield EN3 7SJ
Tel: 01-804-1378

This company distributes the snap camera which was described in the section on pattern sensing. They also market a range of robotic and other peripheral equipment for computers.

For more addresses, see the excellent directory published by Morgan-Grampian which is mentioned in the Bibliography.

# Bibliography

*The Art of Micro Design*
A.A. Berk
Published by Newnes-Butterworths

This book is intended to provide the full details of the design of micro-based controllers and peripheral devices from scratch. It provides industrial examples, and actual circuits, of the way in which MPUs and their peripheral circuits are designed and used. In many ways, the present book is a precursor to *The Art of Micro Design*, where the electronic details of the process and product control designs mentioned here may be found.

*Microprocessor Interfacing Techniques*
Lesea and Zaks
Published by Sybex

The book was one of the first and most readable treatises on this subject. It is still considered to be an important source book for anyone wishing to use microelectronics and peripherals.

*A Handbook of Industrial Control*
E.A. Parr
Published by Collins Technical Books

This is a three volume work. The first volume provides a very complete and practical introduction to sensors and transducers of all kinds. The other books contain all the details needed for a practising electronic engineer wishing to use control techniques.

There are many books of a general nature on the subject of microprocessors, and two examples are given next which provide a practical approach to the subject.

*Interfacing to Microprocessors and Microcomputers*
Owen Bishop
Published by Newnes Microcomputer Books

This book has full details of projects which can be constructed even by a beginner.


*Introduction to Microprocessor System Design*
Harry Garland
Published by McGraw-Hill

This book is a general and readable introduction to the field of MPUs.


*Transducer Interfacing Handbook*
D.H. Sheingold
Published by Analog Devices Inc.,
Address in UK:
Central Avenue,
East Molesey,
Surrey KT8 0SN

Analog Devices Inc. manufacture an excellent range of integrated circuits for interfacing transducers and sensors. This book provides general and applications information for many of their products.


*Control Engineering*
Noel M. Morris
Published by McGraw-Hill

This is an excellent introduction to control theory and practice for the engineer. Noel Morris has also written a previous book called *Advanced Industrial Electronics*, by the same publisher. This is also highly recommended for source material and data for designing systems.


*Transducers, Sensors and Detectors*
Robert G. Seippel
Published by Prentice-Hall

This book supplies some general information on the types of sensors which are used in modern control, including photographs wherever possible.

*Directory of Electronics Instruments and Computers*
Published by Morgan-Grampian Book Publishing Co Ltd
30 Calderwood St,
London SE18 6QH
Tel. 01-855 7777

This directory contains a very full list of companies involved with all the devices and subjects mentioned in this book. It is sectioned off into subject headings to allow the reader to find a list of companies which deal with any given area of technology.

# INDEX